The Official
ENGLAND
World Cup Book 2002

The Official ENGLAND World Cup Book 2002

Executive Editor Vanessa Daubney
Editorial Justyn Barnes & Aubrey Ganguly
Design David Hicks
Jacket Design Steve Lynn
Picture Research Debora Fioravanti
Production Lisa French
Project Art Direction Mark Lloyd
Senior Art Editor Darren Jordan

Thanks also to Marc Armstrong, David Barber, Ben Wright and Giovanna Palladino at The Football Association for all their help in the production of this book.

The F.A. Crest and F.A. England Crest are official trade marks of The Football Association Limited and are the subject of extensive trade mark registrations worldwide.

THIS IS A CARLTON BOOK
Copyright © Carlton Books Limited 2002

1 3 5 7 9 10 8 6 4 2

This book is sold subject to the condition that it shall not, by way of trade or otherwise, be lent, resold, hired out or otherwise circulated without the publisher's prior written consent in any form of cover or binding other than that in which it is published and without a similar condition, including this condition, being imposed upon the subsequent purchaser.

All rights reserved.

A CIP catalogue for this book is available from the British Library.

ISBN: 1 84222 588 X

Printed and bound in Italy

CONTENTS

Welcome!

Hello and welcome to The Official England World Cup Book 2002.

It's been a tough battle to qualify for Korea/Japan 2002 but we made it in the end. There were times when even I thought we wouldn't qualify automatically... for instance, just before David Beckham scored against Greece! But the character and skill of the players got them through and they thoroughly deserve their chance to play in football's greatest tournament. I'd like to take this opportunity to thank all those who played for us in the qualifiers, everyone at The Football Association and, of course, you, the fans. Your fantastic support is like having an extra player. Now, I can't wait for the tournament to start and I hope we can give you plenty more to cheer about this summer!

Sven-Göran Eriksson

6 GO EAST!
The amazing story of England's World Cup 2002 qualifying campaign

14 CAPTAIN FANTASTIC
England skipper David Beckham looks forward to the big one

16 ENDURING IMAGE No. 1
The first of four unforgettable England World Cup images: Gazza's tears

17 SVEN-GALI
Our favourite Swede reveals his football philosophy

18 NUMBER ONE
From Bert 'The Cat' Williams to David Seaman... a tribute to England's World Cup 'keepers

20 RESPECT DUE No. 1
A big hand for six of the best opposition players: First, it's French genius Zinedine Zidane

21 WELCOME TO THE FUTURE
Under–21 coach David Platt on England's best youth prospects

22 THE FINEST
He's the European Footballer Of The Year and now Michael Owen craves World Cup glory

24 ADAM CROZIER
The F.A. Chief Executive on appointing Sven and supporting England

25 RESPECT DUE No. 2
Portuguese maestro Luis Figo

26 RIO, WES & SOL
A closer look at a trio of classy England defenders

28 THE MAN WHO KNOWS
Italia '90 England boss Bobby Robson on the pressure and pleasure of a World Cup tournament

30 RESPECT DUE No. 3
A round of applause for Rivaldo

31 RIGHT-HAND MAN
Meet Sven's trusted confidante Tord Grip

32 HERE WE GO!
Argentina, Nigeria, Sweden... your insider's guide to England's Group F opponents

36 THE ENGINE ROOM
Liverpool's non-stop midfielder Steven Gerrard gears up for Japan 2002

38 ENDURING IMAGE No. 2
Was it or wasn't it? Geoff Hurst's controversial goal in the 1966 World Cup Final

40 HANDLE WITH CARE
Physio Gary Lewin explains how he helps to get England's stars ready to play

41 RESPECT DUE No. 4
All hail 'The Angel' Gabriel Batistuta

42 PARTY TIME!
Your comprehensive guide to staging a themed World Cup bash for your mates!

44 BIG GUNS
In-depth profiles of favourites Brazil, France, Italy and Spain.

48 WORLD CUP FIXTURES
Keep the scores of every World Cup game using our tournament planner

50 DARK HORSES
Will Cameroon, Croatia, Germany or Portugal spring a surprise?

54 REST OF THE WORLD
Scouting reports on all the other World Cup contenders from Belgium to Uruguay

64 THE GAFFER
An exclusive interview with England manager Sven-Göran Eriksson

66 EAT TO WIN!
Gourmet chef Roger Narbett on feeding England's élite squad

67 RESPECT DUE No. 5
Irishman Roy Keane takes 'em all on

68 MIDFIELD MAGICIAN
We talk to Paul Scholes. He scores goals, you know...

70 ENDURING IMAGE No. 3
1982: Robbo scores one of the fastest goals in World Cup finals history

72 THE KNOWLEDGE
Scare your friends with your grasp of England World Cup trivia!

73 RESPECT DUE No. 6
Francesco Totti – Sir Alex Ferguson says he's the best player in the world

74 THE QUEST FOR GLORY
Re-live the ups and downs of England's World Cup history

82 FROM SAITAMA TO YOKOHAMA
Handy travel guide for England fans going all the way to Japan 2002

86 ENDURING IMAGE No. 4
Michael Owen's wonder goal v Argentina at France '98

88 WORLD IN MOTION
Get down and boogie to, ahem, magical England retro grooves

89 WORLD CUP FUNNIES
Rib-tickling World Cup cartoons by *Private Eye's* Tony Husband

90 THE LIKELY LADS
Up close and personal with the England wannabes hoping to make Sven's final 22

96 SNAPPED
Picture credits

ROAD TO WORLD CUP 2002

GO EAST!

It was one hell of a ride but we got there in the end! Let's re-live the lows, the highs and the nail-biting tension of England's tumultuous World Cup 2002 qualifying campaign...

7 OCTOBER 2000, WEMBLEY

ENGLAND 0 GERMANY 1
HAMMAN 14

BOSS KEVIN KEEGAN WAS UNDER intense pressure in the build-up to our first 2002 World Cup qualifier, and the last international ever to be played at the old Wembley. After England's failure to qualify from the group phase at Euro 2000, everyone seemed to have an opinion on what was wrong with the national team. Even celebrity cook Nigella Lawson took it upon herself to offer a diagnosis – 'The lack of PE classes in schools has made us a fat and indolent nation... the pool of strong, healthy English people simply isn't big enough,' claimed the kitchen goddess. Hmm. More seriously, football journalists and fans were questioning Keegan's tactical ability, and his choice to play Gareth Southgate in midfield only added fuel to the fire. Keegan refused to be depressed by the prevailing mood of gloom though. 'You can doubt me, you can question my suitability for the job, but my life's been about proving people wrong,' he stated defiantly.

England had beaten a poor German side 1–0 in Charleroi at Euro 2000, our first competitive win over our fierce rivals since 1966, and earned a creditable 1–1 draw in a friendly against France since the tournament. But any optimism was tempered by Germany's two encouraging wins under new manager Rudi Völler and their determination to gain revenge. 'We feel there is some pride to win back,' said German midfielder Dietmar Hamann in a pre-match interview.

The game was played in shocking conditions – leaden skies, pouring rain and a sodden pitch – and England's performance was just as miserable. From the start, it was obvious that our German opponents were a different proposition to the team that had capitulated meekly in Charleroi. They were fighting for every ball and Mehmet Scholl in particular was causing problems with his

Seaman repels another German attack

skill and intelligent running.

The decisive moment of the match occurred on 14 minutes. There appeared to be little danger when Paul Scholes fouled Michael Ballack 25 yards out, but while 'keeper David Seaman was organising the defensive wall, quick-thinking Hamann fired a low free-kick that beat Seaman's late dive.

At half-time, Keegan brought on Dyer for Gary Neville, moved Southgate back into defence and switched from a 4–1–3–2 to a 3–5–2 formation. But although Dyer added attacking thrust, England were largely limited to taking potshots from long-range as the Germans defended resolutely and threatened intermittently on the break.

So England's first World Cup qualifier ended in defeat and the team trudged off to a cacaphony of boos. In the tunnel, an emotional Kevin Keegan called time on his England managerial career. It was a sad finale for this likeable man and also for a world famous 77-year-old stadium.

> 'In the Wembley tunnel, an emotional Kevin Keegan called time on his England managerial career.'

Teddy can't break the deadlock

11 OCTOBER 2000, HELSINKI

FINLAND 0 ENGLAND 0

FOLLOWING KEVIN KEEGAN'S SHOCK resignation, F.A. Technical Director Howard Wilkinson took the responsibility of overseeing England's Helsinki trip. He called in Leeds United's Director Of Youth Brian Kidd and legendary England warhorse Stuart Pearce to help with the coaching and recalled evergreen striker Teddy Sheringham to the playing squad. With just a couple of days to prepare, Wilkinson's cause wasn't helped by injuries to key players – David Beckham (knee) and Steven Gerrard (thigh) were both sent home after the Germany game. But Wilko called upon the remaining players to 'show pride' adding that 'the current perception of the players doesn't mirror their capabilities'.

But the players seemed unsettled by the Germany reverse and the sudden departure of their manager, and played nowhere near their potential. Luck was against them too in two crucial incidents. Firstly, Finnish goalkeeper Antti Niemi should have been sent off for upending Teddy Sheringham but escaped with a mere booking. Then, in the closing minutes of the match, Ray Parlour ran from midfield before hitting a shot against the crossbar. Replays showed the ball bounced down over the line, but the officials waved play on.

The goalless draw, combined with Albania's 2–0 victory over Greece, left England two points adrift at the bottom of Group Nine. Anyone for the England manager's job?

ENGLAND LINE UP
(4–1–3–2 FORMATION)

D SEAMAN
G NEVILLE (K DYER 46),
T ADAMS
M KEOWN
G LE SAUX (G BARRY 77)
G SOUTHGATE
D BECKHAM (R PARLOUR 82),
P SCHOLES
N BARMBY
ANDREW COLE
M OWEN

SUBS NOT USED
N MARTYN
E HESKEY
D WISE
K PHILLIPS
BOOKED
ANDREW COLE

	P	W	D	L	F	A	GD	Pts
GERMANY	2	2	0	0	3	0	+3	6
FINLAND	2	1	0	1	2	2	0	3
GREECE	2	1	0	1	1	2	-1	3
ALBANIA	1	0	0	1	1	2	-1	0
ENGLAND	1	0	0	1	0	1	-1	0

ENGLAND LINE UP
(4–4–2 FORMATION)

D SEAMAN
P NEVILLE
M KEOWN
G SOUTHGATE
G BARRY
(W BROWN 69)
R PARLOUR
D WISE
P SCHOLES
E HESKEY
E SHERINGHAM
(S McMANAMAN 69)
ANDREW COLE

SUBS NOT USED
N MARTYN, N BARMBY
R FERDINAND, J COLE
M OWEN

	P	W	D	L	F	A	GD	Pts
GERMANY	2	2	0	0	3	0	+3	6
FINLAND	3	1	1	1	2	2	0	4
ALBANIA	2	1	0	1	3	2	+1	3
GREECE	3	1	0	2	1	4	-3	3
ENGLAND	2	0	1	1	0	1	-1	1

ROAD TO WORLD CUP 2002

24 MARCH 2001, ANFIELD

ENGLAND 2   FINLAND 1
OWEN 42, BECKHAM 49   RIIHILAHTI 27

THE F.A.'S DECISION TO EMPLOY A foreign manager was controversial, but critics who slammed Sven-Göran Eriksson's appointment began to quieten after this 2–1 win, England's first competitive game under the Swede.

The previous month, Sven's side had beaten Spain 3–0 in a friendly which showed that the players were already responding to the Swede's calm approach. So it was with a sense of renewed confidence that fans and players approached this crucial qualifier at Anfield. The team showed eight changes from Howard Wilkinson's emergency XI that drew 0–0 in Helsinki, including 31-year-old Chris Powell, whom Eriksson had brought in to fill the problematic left-back position against Spain, and Michael Owen who would operate just behind Andrew Cole up front.

England started promisingly, dropping into a nice passing rhythm on the slick Anfield pitch. But as the first half wore on, the link between England's midfield and attack weakened and the passes became longer and more hopeful. Then, on 27 minutes, the visitors took a shock lead, when Aki Riihilahti was allowed a free header, and Gary Neville's knee deflected the ball past Seaman.

But just when it seemed Eriksson's half-time team talk would be a gloomy one, his captain David Beckham showed the determination that earned him the armband. Becks ran at the Finnish defence from midfield before feeding the ball to Gary Neville on the right. His cross flashed across the area to Michael Owen who fired in the equaliser.

In the 49th minute, it was Beckham's turn to finish off a fine team move, the Manchester United star's unstoppable swerver making it 2–1 and earning him an unprecedented ovation from the Kop. England missed further opportunities to make the game safe, and almost paid for their profligacy when Jari Litmanen seemed certain to equalise only to be denied by Seaman's point-blank save.

Not a perfect performance then, but a lot better and, most importantly, England had three more points in the bag.

ENGLAND LINE UP
(4-4-2 FORMATION)

D SEAMAN
G NEVILLE
S CAMPBELL
R FERDINAND
C POWELL
D BECKHAM
S GERRARD
P SCHOLES
S McMANAMAN
(E HESKEY 71)
ANDREW COLE
(R FOWLER 81)
M OWEN
(N BUTT 89)

SUBS NOT USED
P NEVILLE, N MARTYN
W BROWN, E SHERINGHAM
BOOKED S McMANAMAN, P SCHOLES

	P	W	D	L	F	A	GD	Pts
GERMANY	3	3	0	0	5	1	+4	9
ENGLAND	3	1	1	1	2	2	0	4
FINLAND	4	1	1	2	3	4	-1	4
ALBANIA	3	1	0	2	4	4	0	3
GREECE	3	1	0	2	1	4	-3	3

Owen fires home the equaliser

Cole clinches victory with his first international goal

28 MARCH 2001, TIRANA

ALBANIA 1 ENGLAND 3

RRAKLI 90   OWEN 74, SCHOLES 85, COLE 90

AGEING ENGLISH COMIC NORMAN Wisdom is a national hero in Albania, but England weren't laughing until the very end of this tense match in the country's capital.

The first-half had seen few chances created on either side, with only England captain David Beckham imposing himself on the game and going close with two free-kicks. Emile Heskey came on as substitute at half-time for Steve McManaman and made his presence felt, but the battling Albanians continued to frustrate Eriksson's men.

Nicky Butt was denied a blatant penalty in the 62nd minute and it was another 12 minutes before England's persistence finally paid off. Paul Scholes' beautifully-weighted first-time pass from midfield split the defence and allowed Michael Owen to slot the ball under the goalkeeper. It was Owen's seventh away goal for England and it lifted the team, but a second didn't come until just after Owen was substituted. This time, much-maligned striker Andrew Cole was the creator. He beat the Albanian left-back and drove into the six-yard box before pulling the ball back into the path of his Manchester United team-mate Paul Scholes who gratefully smashed the ball into the roof of the net.

Scholes' 11th international goal should have sealed the win, but two minutes into stoppage time indecision in the England defence let Albania substitute Altin Rrakli in to score. Even worse, the Albanians had the ball in the net again a minute later, only for the goal to be (wrongly) ruled offside. Still there was more injury time to play and England decided attack was the best way to defend their precarious lead. And with the referee about to blow for full time, Emile Heskey broke down the left wing before cutting the ball back to Andrew Cole who converted his first international goal. It was rich reward for Cole who'd been subjected to vicious media criticism since the Finland game when his good performance was marred by one missed chance.

Afterwards, a delighted Sven-Göran Eriksson commented, 'Andy Cole played very well, and so did the rest of the team. We gave away almost nothing to Albania. Now we have to go on from here.'

ENGLAND LINE UP
(4–4–2 FORMATION)

D SEAMAN
G NEVILLE
S CAMPBELL
(W BROWN 29),
R FERDINAND
ASHLEY COLE
D BECKHAM
N BUTT
P SCHOLES
S McMANAMAN
(E HESKEY 46)
ANDREW COLE
M OWEN
(E SHERINGHAM 84)

SUBS NOT USED
P NEVILLE, N MARTYN,
J CARRAGHER, R FOWLER
BOOKED G NEVILLE, ANDREW COLE

	P	W	D	L	F	A	GD	Pts
GERMANY	4	4	0	0	9	3	+6	12
ENGLAND	4	2	1	1	5	3	+2	7
FINLAND	4	1	1	2	3	4	-1	4
ALBANIA	4	1	0	3	5	7	-2	3
GREECE	4	1	0	3	3	8	-5	3

ROAD TO WORLD CUP 2002

6 JUNE 2001, ATHENS

GREECE 0 ENGLAND 2
SCHOLES 64, BECKHAM 87

FOLLOWING AN IMPRESSIVE 4–0 FRIENDLY win over Mexico in May, England extended their winning streak to five games in this World Cup qualifier in Athens. The 2–0 victory meant Sven-Göran Eriksson had made the best start of any England coach in history.

Any concerns that fatigue might affect players reaching the end of the long season were dismissed as England produced another assured display. In a solid but unadventurous first half, striking duo Robbie Fowler and Michael Owen offered England's best chance of a breakthrough. Fowler went close with a bullet header on 32 minutes before setting up Owen with a great pass from inside his own half. Uncharacteristically, Owen could only find the side netting with his shot, but the Liverpool team-mates were showing promise as a partnership.

In the second half, captain David Beckham's skill and tireless endeavour inspired those around him. And on 64 minutes England took the lead after a fine team move. An intuitive reverse pass by Fowler set Phil Neville free. Neville took the ball past two challenges and fed it to Emile Heskey who vindicated Sven-Göran Eriksson's decision to play him on the left with a delightful cross that Paul Scholes powered home (his third goal in three games).

England were rarely troubled after that and it was fitting that the magnificent Beckham added a late second goal with a sublime 28-yard free-kick.

'It was a great team performance,' said Beckham afterwards. 'We worked hard and got the result we wanted.'

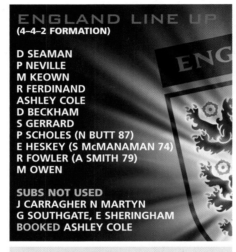

ENGLAND LINE UP
(4–4–2 FORMATION)

D SEAMAN
P NEVILLE
M KEOWN
R FERDINAND
ASHLEY COLE
D BECKHAM
S GERRARD
P SCHOLES (N BUTT 87)
E HESKEY (S McMANAMAN 74)
R FOWLER (A SMITH 79)
M OWEN

SUBS NOT USED
J CARRAGHER N MARTYN
G SOUTHGATE, E SHERINGHAM
BOOKED ASHLEY COLE

	P	W	D	L	F	A	GD	Pts
GERMANY	6	5	1	0	13	5	+8	16
ENGLAND	5	3	1	1	7	3	+4	10
GREECE	6	2	0	4	4	10	-6	6
FINLAND	5	1	2	2	5	6	-1	5
ALBANIA	6	1	0	5	5	10	-5	3

Scholes strikes again!

WHAT A NIGHT! THIS HISTORIC win completed England's remarkable transformation from Group Nine no-hopers after two games to favourites with two games to play. It was a wonderful performance, especially considering that just two weeks earlier England had been beaten 2–0 by Holland in a friendly, the team's first defeat under Sven-Göran Eriksson.

England fans got a boost before a ball was kicked when captain David Beckham led the team out at the Olympic Stadium. Beckham had been suffering from a knee injury all

Steve Gerrard 'swans' off after putting us 2–1 up

1 SEPTEMBER 2001, MUNICH

GERMANY 1 ENGLAND 5
JANCKER 6   OWEN 13, 48, 66, GERRARD 45, HESKEY 73

week, but declared himself fit to play.

However, Germany had the better of the early exchanges and England were grateful to Steven Gerrard (who also passed a late fitness test) for thwarting a threatening run into the area by Sebastian Deisler. Deisler had been touted as Germany's danger man and his chip into the box on six minutes caused panic in the England defence allowing Carsten Jancker to score.

England were back on level terms within seven minutes. Beckham over-hit a free-kick from near the corner flag, but Gerrard diligently chased the ball down and passed it to Gary Neville who played it back into the area. Barmby's clever flicked header wrong-footed the German defence and Michael Owen drilled the ball past Oliver Kahn.

England grew in confidence, with Beckham, Gerrard and Paul Scholes working tirelessly in midfield, but the occasional defensive lapse still gave Germany encouragement. Indeed, Deisler should have restored their lead on 22 minutes but skewed his shot wide. Then, with the half-time interval beckoning, Beckham swung a free-kick towards the edge of the box. Rio Ferdinand headed the ball into the path of Steven Gerrard who struck a sweet half-volley into the corner of the net – 2–1!

After the interval, things just got better and better. Owen notched his second on 48 minutes and just over a quarter of an hour later he raced onto a perfect pass by Gerrard before calmly curling the ball over Kahn's out-stretched hand. It was Owen's 13th goal in 21 international starts and made him the first Englishman since Sir Geoff Hurst to score a hat-trick against Germany.

Seven minutes later, Emile Heskey completed the rout, striking the ball under Kahn's body from Scholes' through-ball.

'5–1 is unbelievable,' said hat-trick hero Michael Owen later. 'No one could have predicted that, but it is about time we showed what we can do.'

Suddenly, Japan and Korea didn't seem so far away after all.

ENGLAND LINE UP
(4–4–2 FORMATION)

D SEAMAN
G NEVILLE
S CAMPBELL
R FERDINAND
ASHLEY COLE
D BECKHAM
P SCHOLES
(J CARRAGHER 84)
S GERRARD
(O HARGREAVES 77)
N BARMBY (S MCMANAMAN 65)
M OWEN
E HESKEY

SUBS NOT USED
G SOUTHGATE, N MARTYN
R FOWLER, ANDREW COLE
BOOKED E HESKEY

	P	W	D	L	F	A	GD	Pts
GERMANY	7	5	1	1	14	10	+4	16
ENGLAND	6	4	1	1	12	4	+8	13
FINLAND	6	2	2	2	7	6	+1	8
GREECE	6	2	0	4	4	10	-6	6
ALBANIA	7	1	0	6	5	12	-7	3

ROAD TO WORLD CUP 2002

5 SEPTEMBER 2001, ST JAMES' PARK

ENGLAND 2 ALBANIA 0
OWEN 44, FOWLER 88

Robbie Fowler scores the second goal for England

AFTER THE HYSTERIA OF MUNICH, IT was important that England didn't undo all their good work four days later. And although the 51,000 England fans who converged on St James' Park may have hoped for another cricket score, they went away revelling in the fact that a workmanlike 2–0 win had taken England to the top of Group Nine.

Still on a high after Saturday's scintillating performance, England began at a quick tempo and might have been awarded a penalty when Michael Owen was felled by Albanian goalkeeper Foto Strakosha after just two minutes. But the game soon fell into a slumber as England struggled to maintain the same levels of energy and fluidity that had overpowered Germany.

England finally took the lead just before the half-time interval. Paul Scholes spotted Michael Owen's run through the middle and his perfectly-weighted chip was clipped past Strakosha by the in-form Liverpool striker.

Thereafter, the mental and physical exertions of Munich took their toll and Albania had the better of the second half – the visitors' best effort was a swerving, dipping shot by Edvin Murati on 77 minutes which David Seaman did well to repel. So it came as some relief when, with two minutes remaining, substitute Robbie Fowler tricked his way past one defender, nutmegged another and clipped the ball over the advancing Strakosha to score a super individual goal and secure a fifth successive World Cup qualifying win.

ENGLAND LINE UP
(4-4-2 FORMATION)

D SEAMAN
G NEVILLE
S CAMPBELL
R FERDINAND
ASHLEY COLE
D BECKHAM
S GERRARD (J CARRAGHER 81)
P SCHOLES
N BARMBY
(S McMANAMAN 62)
M OWEN
E HESKEY (FOWLER 53)

SUBS NOT USED
G SOUTHGATE, R WRIGHT
O HARGREAVES, ANDREW COLE
BOOKED S GERRARD

	P	W	D	L	F	A	GD	Pts
ENGLAND	7	5	1	1	14	4	+10	16
GERMANY	7	5	1	1	14	10	+4	16
FINLAND	7	3	2	2	12	7	+5	11
GREECE	7	2	0	5	5	15	-10	6
ALBANIA	8	1	0	7	5	14	-9	3

ALL WE NEEDED WAS A NICE HOME win against Greece, a team that had scored just one goal away from home in Group Nine, and we'd be on our way to Korea/Japan 2002. What could possibly go wrong? Well, just about everything until David Beckham's miraculous last-gasp equaliser.

Manager Sven-Göran Eriksson later admitted that 'it wasn't an easy game to watch', and 66,090 people inside Old Trafford plus millions of England fans watching on television certainly agreed.

Beckham apart, England looked edgy from the kick-off. Nigel Martyn had already been called upon to make sharp saves from Theo Zagorakis and Angelos Charisteas, so it was no great surprise when Greece took the lead on 36 minutes. Christos Patsatzoglou

ENGLAND LINE UP
(4-4-2 FORMATION)

N MARTYN
G NEVILLE
M KEOWN
R FERDINAND
ASHLEY COLE
(S McMANAMAN 78)
D BECKHAM
P SCHOLES
S GERRARD
N BARMBY
(ANDREW COLE 46)
E HESKEY
R FOWLER
(E SHERINGHAM 68)

SUBS NOT USED
R WRIGHT, G SOUTHGATE
J CARRAGHER, D MURPHY
BOOKED P SCHOLES

GROUP 9 FINAL TABLE

	P	W	D	L	F	A	GD	Pts
ENGLAND	8	5	2	1	16	6	+10	17
GERMANY	8	5	2	1	14	10	+4	17
FINLAND	8	3	3	2	12	7	+5	12
GREECE	8	2	1	5	7	17	-10	7
ALBANIA	8	1	0	7	5	14	-9	3

rounded Ashley Cole on the right, and Rio Ferdinand's headed clearance fell at the feet of Charisteas who powered an unstoppable shot into the corner of the net.

At half-time, 1–0 down, Sven-Göran Eriksson made a tactical change bringing on Andrew Cole for Nicky Barmby and switching Emile Heskey to the left wing so Cole could play up front with Fowler. Cole almost scored with his first touch, latching on to Gerrard's cute chip but his shot was smothered by Greek 'keeper Antonis Nikopolidis. Three minutes later, England had a scare when Giorgos Karagounis broke free, but Nigel Martyn made a vital save.

Amid mounting tension, England won a free-kick in the 68th minute and Eriksson took the opportunity to make his second substitution replacing Robbie Fowler with Teddy Sheringham. Sheringham jogged on as Beckham prepared to take it and brought new meaning to the term 'timing your run into the box' by meeting the captain's cross with a deft header to score with his first touch.

The joy lasted less than a minute however as Themis Nikolaidis scored from close range to restore Greece's advantage. England

continued to probe but to no avail and were grateful when referee Mr Jol added on four minutes of stoppage time. After three of them, Sheringham won a free-kick 25 yards out. Beckham had missed from a similar spot 10 minutes earlier, but under the most intense pressure he unleashed a beautiful curling shot that hit the net before Nikopolidis could move a muscle. Players, fans and staff went crazy and the news that Germany had only managed a home draw against Finland meant England's place in the World Cup Finals was sealed.

'It was a great day for football,' said an ecstatic Sven-Göran Eriksson later, 'not necessarily in the way we played – because we have certainly played better – but in the drama of the match. I'm delighted to have qualified.' So are we, Sven. So are we.

6 OCTOBER 2001, OLD TRAFFORD
ENGLAND 2 GREECE 2
SHERINGHAM 68, BECKHAM 90   CHARISTEAS 36, NIKOLAIDIS 69

'Players, fans and staff went crazy and the news that Germany had only managed a draw meant England's place in the World Cup Finals was sealed.'

LIONHEARTS

CAPTAIN FANTASTIC

After France '98 DAVID BECKHAM was Public Enemy No. 1, but now he is a national hero and Becks can't wait to lead his country at the World Cup...

'We all have dreams but to score a free-kick in the dying seconds was unbelievable.'

HERE COMES DAVID BECKHAM. ONE half of 'Britain's most stylish couple' according to *OK!* magazine. A founding member of Posh and Becks Inc – the country's biggest small company reckons *Heat*. A 'celebrity doll twisted and pulled out of all recognition' claims *Esquire*. How can anyone cope with that much attention?

But I digress. Here comes David Beckham, squinting in the early morning sun, just about to start training. Not looking particularly stylish if truth be told, in sweatshirt and shorts, socks rolled down to his ankles. Nor bearing much resemblance to a small corporation. As for a celebrity doll... well, maybe Action Man when it boasted 'real hair'.

The truth is David Beckham simply looks like a footballer, especially since he lost the pop star locks. And it's as a footballer that he matters to England. How the Manchester United midfielder performs in Japan and Korea will probably have the biggest bearing on how far the national team progresses in this summer's World Cup Finals.

'David Beckham is England's main player,' Roberto Carlos once claimed. 'He can cause any team in the world problems.'

Such observations could leave a lesser character feeling under pressure to perform but Beckham seems to relish the challenge.

'I can't wait for the tournament to begin,' he grins. 'As a footballer you always want to test yourself against the best. And with the draw England have got, there's no doubt we'll get to do that. We're facing three very different teams in the first stage, but what they have in common is that they're all very good.'

Sweden and Nigeria are certainly good teams but the biggest challenge is surely provided by Argentina, a team the England captain knows all about. 'Yes, everybody knows there is some history between the teams,' says Beckham. 'But this is a great chance to put the record straight and ease the memories of what happened in 1998.'

Of course, that year saw Beckham at his lowest point. Sent off in St Etienne and turned into Public Enemy No. 1 by the tabloids, he could have been forgiven for letting his head drop or hitting out at the critics. Instead he let his football do the talking. When this World Cup starts, he will begin it as captain of his country and a hero to millions back home. Funny how things turn out.

'Certain things happen in life that you regret,' reflects David. 'But those things have helped me become a stronger person and in a weird way they were a good thing because of that.'

The final burial of the past came at Old Trafford, when deep into injury time, Beckham struck the free-kick that put England in the finals.

'The most important goal of my career to date' is how he describes it. 'We all have dreams but to score a free-kick in the dying seconds was unbelievable.'

Listening to him talk about football reminds you just what drives David Beckham on. And it's not haircuts or sarongs.

'All I ever wanted was to be a professional footballer. I've never thought of myself as someone special. My family was an ordinary working-class family. I've worked very hard to achieve what I have and I'll carry on working hard to improve my game.'

Ask him when he first realised he could make it as a player and he ponders for a second before breaking into a smile.

'The only way I knew I was good is when I used to go home covered in bruises, from where people used to kick me! Not a lot has changed! People said things about me, but I wasn't a flashy kid. If I was, my dad would have put me back in my place! That's not the kind of person I was.'

Or is. The headlines, truth and fiction, will continue to come. But right now there's only one thing concerning the captain who will be 27 in May.

'We wanted to go to the World Cup finals and now we're there. It was great to play in the last one whatever anyone thinks or says. And now we've got another chance to show what we can do.'

England might just take this chance, if the captain gets his way.

GAZZA'S TEARS

**4 July 1990, Turin
World Cup Semi-Final
England 1 West Germany 1
(AET: West Germany won
4–3 on penalties)**

Terry Butcher consoles Paul Gascoigne after England's heart-breaking defeat on penalties. The tears had started to flow during extra time when Gazza received a booking that would have ruled him out of the final. Sadly, the team didn't quite make it anyway.

SVENGALI

Sven-Göran Eriksson's cool coaching philosophy has turned England from no-hopers into true World Cup contenders. Here's a few of the ways he's helped the lads reach a state of Sven...

Discipline with freedom

'I used to be hard as nails on tactical discipline, but now I go along with the "discipline with freedom" line. The best players in particular can't be put in a straitjacket or they won't be free to use their ability. If I force a player to carry out a task in a certain way, he'll feel like he is playing under compulsion instead of with complete freedom. You can generally say that the longer you have a team together, the greater freedom you can allow them.'

Person & performance

'It is usual, and understandable, for us to connect our own person with our performance. But I believe in just the opposite: separate the person from the performance. Our performance is always going to vary – nobody can always be on top of their form.

'The problem with connecting the performance to the person is that our view of ourselves will go up and down with our performance. With mental training, we can have control over ourselves and our performance, instead of becoming a victim of our performance.'

Team selection

'The first condition for selection of course, is that they should be good players who fit in with our style of play. But I always like to have the right personality types. I like to have leaders. Leader figures have the advantage of being able to speak up and football is also about communication. But you can be a good player without being a leader – you need both leader types and non-leaders to get a good team together.'

Fighting spirit

'Some players and coaches think of "fight" and "aggression" as more or less the same thing, but for me they are entirely different. I see aggression as a "flight" reaction – we feel so much pressure that we respond in panic. Aggression is based on insecurity. Real fighting spirit is based on inner security and self-confidence.

'Aggression blocks a number of areas such as technique, tactics, concentration and assessment. I spend a lot of energy taking the aggression out of my players. All a player has to do is argue with the referee, play dirty or quarrel with the opposition for there to be a danger that their performance level will sink like stones – not only the player's performance but the whole team's.'

Pressure points

'When you are trailing in a league table you have to continue to motivate your players, to calm things down, while you explain that there is still a chance of winning. It is also extremely important to spend extra time talking to my most important players, the ones that take the whole team with them. In pressure situations, you have to expect the younger players to hang back, reacting through fear and simple cowardice. As a coach you can't put too much pressure on them.'

Build an atmosphere

'What is decisive in critical situations is the atmosphere in the squad. But a good atmosphere is not something that can be built in a short time, it takes time and patience. I've always tried to instil a 'we' feeling among the players. I believe the players' mental attitude is even more important during the final stages of a season or a tournament, because by then it is too late to change your tactics or technique.'

Psychology... and luck

'So little is required to be successful in sport. It's certainly mostly a matter of psychology, and ultimately it's that psychological difference that decides whether you win or lose. And, of course, you sometimes need a little luck. Then the impossible can become a reality.'

Taken from 'Sven-Göran Eriksson on Football' with Willi Railo & Håkan Matson, published by Carlton Books

THE ONION BAG

NUMBER 1

From Bert 'The Cat' Williams to David Seaman, let's remember the brave souls who've guarded the onion bag for England in football's biggest tournament...

WHEN ENGLAND TRAVELLED to their first World Cup in Brazil in 1950, Wolverhampton Wanderers goalkeeper **Bert 'The Cat' Williams** was our last line of defence, but the tournament was a bit of a nightmare for the 30-year-old former RAF World War Two pilot. 'The Cat' found himself clawing the ball out of the net against minnows USA (0–1) and Spain (0–1) as England tumbled out of the tournament in the first phase.

Another Midlander, Birmingham City 'keeper **Gil Merrick**, was manager Walter Winterbottom's first-choice 'keeper for the 1954 World Cup. Winterbottom kept faith with Merrick even though he'd been in goal when Hungary destroyed England 7–1 in the run up to the finals. Sadly, 32-year-old Merrick was beaten four times in England's first match in Swizerland (a 4–4 thriller against Belgium) and let in another four in the 2–4 quarter-final defeat by Uruguay. He never played for England again.

When it was time to name his squad for the 1958 World Cup finals in Sweden, Winterbottom opted for **Colin McDonald** (Burnley) and **Eddie Hopkinson** (Bolton) as his two 'keepers. Twenty-seven-year-old McDonald got the nod and played in all four England games. His heroic performance in a the goalless draw against Brazil was among the greatest ever by an England 'keeper and McDonald thoroughly deserved his Goalkeeper Of the Tournament award. Tragically, the following season McDonald sustained a broken leg which ended his career.

The goalkeeping jersey was handed to Sheffield Wednesday's **Ron Springett** for the next World Cup campaign. Springett did well in qualifying, conceding just two goals in four games. But at Chile '62 his weakness against long-range shooting was ruthlessly exposed by defending champions Brazil in the quarter-final. First, he failed to hold Garrincha's free-kick allowing Vava to score and then Garrincha beat Springett with another venomous shot from open play to complete a 3–1 Brazilian victory.

By the time England hosted the 1966 finals, there was another net custodian wearing the Three Lions – Leicester City's **Gordon Banks**. The quiet 29-year-old from Sheffield was an unobtrusive character off the pitch, but a giant on it. Banks kept a clean sheet for 442 minutes during the tournament (Eusebio eventually slotted a penalty past him after 82 minutes of the semi-final) and was a key figure in England's greatest-ever triumph.

But for all heroics in '66, Banks' status as one of the greatest goalkeepers of all time was secured by his impossible save from Pele at the next World Cup in Mexico. Unfortunately, Banks was struck down by a stomach complaint before England's quarter-final against West Germany, and Chelsea keeper **Peter Bonetti** was called up to play at an hour's notice. Like Bert Williams, Bonetti was known as 'The Cat' for his feline reflexes, but he was culpable for two German goals as they came back from 2–0 down to win 3–2 after extra time. It proved to

'Colin McDonald's performance in the 0–0 draw against Brazil in 1958 earned him the Goalkeeper Of The Tournament award.'

be Bonetti's seventh and last international appearance.

Then it was the era of **Peter Shilton**. Ironically, Shilts conceded a similarly 'soft' goal to the tame effort Bonetti let in from Franz Beckenbauer in the 1–1 qualifying draw against Poland which ultimately ended England's hopes of making the 1974 finals. But he went on to become England's most capped player with 125 caps and represent England at three World Cups. He clocked up a then world record of 500 minutes without conceding a goal in the 1982 and 1986 finals. And, incredibly, at the ripe old age of 39, Shilton kept six clean sheets out of six in the 1990 qualifying campaign before playing a key role in England's run to the semi-finals in Italy.

With Shilts retired, Sheffield Wednesday's **Chris Woods** stepped into the breach from the start of the ill-fated 1994 qualifying campaign, but he was later succeeded by Arsenal 'keeper **David Seaman**. The ever-reliable Seaman proved to be Shilton's true successor and, after a storming Euro '96 was an ever-present for England at France '98. Now, the 38-year-old is hoping to wear the Number One shirt again at World Cup 2002... but **Nigel Martyn**, **David James** and **Richard Wright** have other ideas!

RESPECT DUE

ZINEDINE ZIDANE — FRANCE

The 1998 and 2000 World Footballer Of The Year, 'Zizou' is the outstanding talent in a star-studded French squad. His sublime ball skills, creative passing and goalscoring ability will be a key factor as the World Champions bid to retain their title.

WELCOME TO THE FUTURE!

The 2002 England World Cup squad will be packed with gifted young players and Under-21 coach DAVID PLATT says there's plenty more where they came from...

DAVID PLATT WALKS OFF THE PITCH FOLLOWING another vigorous training session. The 35-year-old former England captain who scored 27 goals in a distinguished 62-match international career is wearing the smile of someone who truly loves his work. That is, to help the next generation of players emulate his success and it's a role he clearly relishes.

'I was very excited when I was offered this job,' he enthuses as we settle down on a log bench behind one of the goals to chat. 'I couldn't pass up the opportunity to work with the best young players in the country.'

Since his appointment as full-time coach last July, David has guided England to the Under-21 European Championship finals. Three wins in their last three group games saw them top their qualifying group and a hard-fought 3–2 aggregate play-off win over Holland secured a berth in the eight-team tournament to be contested in Switzerland between 16–28 May.

England have won the tournament twice before in 1982 and 1984, so how does the coach rate our chances of success this time.

'I believe we can win it,' he says confidently. 'I think it's a great sign that I will have trouble narrowing my squad down to 20 for the finals. I named a squad of 25 for the play-off games against Holland and it will be a real headache down-sizing... of course it's a very nice problem to have.'

But is team success or individual player development most important at this level?

'Developing players is paramount but, to me, winning provides the best form of development,' he rationalises. 'The further we go in Switzerland, the more the players will learn about the intensity and pressure of tournament football. Most of them won't have experienced it before and it's totally different. It will be an excellent grounding for the players who want to break into the senior squad for Euro 2004 or the 2006 World Cup finals.'

Recent graduates from Under-21s to the senior squad include Michael Owen, Rio Ferdinand, and Steven Gerrard and Platt believes there are a number of players in his squad who already have the skills to make the leap. When pushed to nominate individuals, he mentions names such as his captain David Dunn, Chelsea's John Terry and Aston Villa striker Darius Vassell, but he adds a rider.

'There are probably four or five with the technical and mental ability to make the senior squad now, but the youthfulness and strength of Sven's squad is such that there are probably only a couple of places available at the moment. If that.'

So how closely does Sven-Göran Eriksson follow the progress of the Under-21s.

'Very closely indeed,' asserts David. 'He speaks to me regularly. Like me, he's committed to turning young talented players into senior internationals. The door is open for all the players if they have the ability and apply themselves properly.'

For more information on England Under-21s visit the F.A. website at **www.TheFA.com**

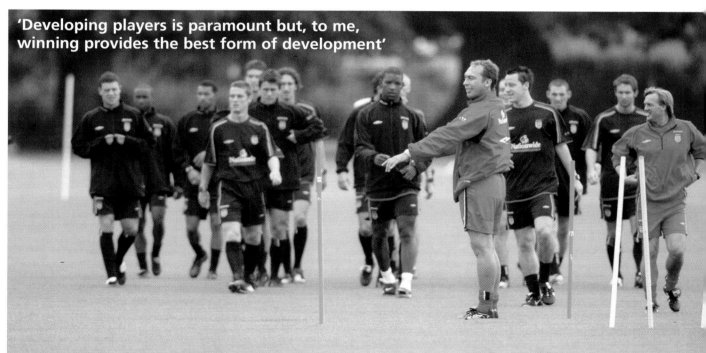

'Developing players is paramount but, to me, winning provides the best form of development'

LIONHEARTS

THE GOAL MACHINE

MICHAEL OWEN became a global star by scoring *that* goal against Argentina at France '98 and he's keen to show it was no fluke...

'The way we have performed over the last year, we have nothing to fear'

STUCK ON A RAINY AFTERNOON IN THE middle of London's rush hour traffic is not everybody's idea of fun. But me? I've no complaints. After all, it gives me more time to interrogate Michael Owen. To be fair, the England striker probably thought we'd only have to chat for about five minutes when he offered me a lift from W1 to W2 but he hadn't allowed for the city's overcrowded roads. Now, it looks like he'll be stuck with this particular journalist for ages. Fortunately, he takes it in good spirit, possibly because we're talking about one of his favourite subjects – the World Cup.

Following the draw, we all know how difficult the task that lies ahead is for England. But the striker is adamant that the nation could have cause to celebrate.

'It is going to be difficult,' he admits. 'We couldn't have picked a tougher group. But the way we have performed over the last year we don't have anything to fear.'

Certainly, there's a growing belief in the country about this England team. Which is in no small part due to Owen's stunning hat-trick against Germany in the qualifying stages. The result made the rest of the world sit up and defenders everywhere slump in their seats. Sven-Göran Eriksson has no doubts about the potency of his striker.

'Michael is something precious and deserves to be mentioned in any list of leading players,' he states. 'With him it is pace, pace and then more pace. No defender, not even the best in the world can live with that forever. He is remorseless. Defenders have to be on their guard the whole time – they cannot relax for a moment.'

Hopefully, the defenders of Nigeria, Sweden and Argentina are already losing sleep. The Argentinians, of course already know what he is capable of. His strike during the 1998 World Cup lit up the tournament and made Owen world famous.

'My life did change,' admits Michael. 'I guess it happened after the World Cup. I was known in this country before but that kind of made the world stand up and take note really. I got a lot more fan mail and a lot more people recognised me. Even when you want to go for something to eat or you want to go shopping, you're constantly being pestered but it's not in any way a criticism because I wouldn't swap my job for anything else.'

And how does he feel about facing the South Americans again, especially now they're many people's tournament favourites?

'I'm quite happy about that. Maybe this is the chance to prove that goal wasn't a fluke and I'll be able to get another just like it.'

Whatever happens in Japan and Korea, Owen has already written his name in the football history books. He became the youngest player of the twentieth century to receive an England cap aged just 18 years and 59 days, (three months younger than the previous record-holder, Duncan Edwards) and immediately offered a statement of intent with a man-of-the-match performance against Chile. Three months later, he scored the first of many goals for England in the 1–0 victory over Morocco. Then followed France '98 and that life-changing goal.

Since then he has tasted lows – injury and a backlash from the build 'em up/knock 'em down tabloids – but has returned to his best form, the former adversities only adding experience to his raw talent. He remains philosophical about the press box's current infatuation with him.

'I prefer the press to be on my side and saying nice things about me rather than bad things but I've experienced both sides. That's just part of the job. If you get injured, you come back and certain people are saying "Is he going to be the player he was?" and things like that. In a way, you have to feel like you've got to prove yourself all over again. You shouldn't really have to.'

Not that Michael has anything left to prove. If the goal against Argentina hadn't already secured his reputation, the hat-trick against Germany certainly did.

'Maybe,' he grins. 'But even when doubts were expressed I didn't feel under pressure because I don't feel I have anything to prove to myself. I know what I'm capable of. If people want me to prove things, they just have to watch me on the pitch because that's where I do a lot of the talking.'

Expect Michael Owen to be very chatty in Japan this summer.

THE CHIEF EXECUTIVE

VIEW FROM THE TOP

Last year was particularly busy for ADAM CROZIER as he oversaw the arrival of a new England manager and watched delightedly as the England team experienced a dramatic change in fortune

IT'S 2.30PM ON A Wednesday afternoon and I've come to the offices of the Football Association in Soho Square in London to chat to The Football Association's Chief Executive Adam Crozier about the events of the past year which saw England defy the gloomy predictions about their chances of qualifying for the World Cup in such style.

I begin by reminding him of the England team's situation back in October 2000. After two World Cup qualifying games, we were goalless, managerless and bottom of UEFA Group Nine. Surely he didn't fancy our chances of qualifying then?

'Well, I thought we would qualify,' he smiles wryly, 'but only through a play-off. The 0–0 draw in Finland was a watershed – after that we said we were going to change the whole structure. And the progress since has been amazing.'

Team England: F.A. Chief Exec Adam Crozier works closely with Sven

A key factor in England's transformation was the appointment of Sven-Göran Eriksson as coach. Initially, despite the Swede's impeccable credentials, there was widespread opposition to the choice of a foreign coach by The F.A.'s selection panel and Scot Crozier bore the brunt of critics' wrath.

But The F.A. had gone to extreme lengths to make sure they got the right man, seeking expert advice from top managers and coaches. Then, once Sven had been targeted, Crozier put great emphasis on speaking to the man himself:

'In the past, I don't think people have put enough thought into talking to the individual about their philosophy, how they would handle the job, who they might want with them etc. We did that with Sven when we first met him back in November 2000 and it was soon clear he was absolutely ideal for the job.'

As England's performances and results improved dramatically under Eriksson, with wins against Finalnd, Albania and Greece, erstwhile doubters among the fans and within the football world were forced to admit they had been proved wrong. But Crozier refuses to gloat:

'I don't feel vindicated for appointing Sven, just really glad it's worked out,' he shrugs. 'From the start, people could see that Sven was working exceptionally hard and that's why he has received a fantastic reception at every ground he's visited.'

Anyone who doubts whether a Scot can be also be a passionate England fan need only ask Crozier for his memories of the turbulent qualifying campaign. Say, beating Germany 5–1.

'That was one the best feelings you could ever experience,' he enthuses. 'I had to leave the directors' box because I just wouldn't sit down! It felt like all of the efforts made by Sven, the coaches, players and The F.A. staff came together that night.'

And what about the agonies of that nail-biting final game against Greece? If he couldn't sit still for the match against Germany how did he find the tension at Old Trafford as he and just about everybody else in the country held their breath in the final moments?

'it was a total nightmare, until the last kick when David Beckham scored,' he groans. 'Afterwards, I was sitting with Sven, Tord, Steve McClaren, Sammy Lee and Dave Sexton – all of us were speechless. We were so happy and relieved that we'd qualified but couldn't believe how it happened.'

So now we've made it to the finals, can we win it? Crozier is cautiously optimistic:

'Because we've got a young side, it's more about continuing improvement. But I think Sven has moved the team forward quicker than even he expected. When we first met Sven, we stated our aim was to win a trophy by 2006 and he felt that was realistic... but that doesn't mean we've ruled out success before then!'

RESPECT DUE

LUIS FIGO — PORTUGAL

As a teenager Luis Felipe Madeira Caeiro Figo won both the Under-16 and Under-20 World Cups and he has high hopes of winning the senior version at Korea/Japan 2002. The gifted right winger was instrumental in Portugal's run to the Euro 2000 semi-finals and he should thrive in the heat of the Far East.

TRIO

Rio, Sol, Wes... three-lettered monikers are not the only thing Ferdinand, Campbell and Brown have in common. The trio are also all capable of performing on the world stage for years to come, which bodes well for the future of England's defence.

IT'S ALWAYS INTERESTING WATCHING PROFESSIONAL footballers train, if not quite as revelatory as you might hope. Apart from their first touch being rather better than yours or mine, there's not a lot of discernible difference between the England squad kicking a ball around and a bunch of mates having a kickaround in the park. As we sit and watch Sven's squad go though their routines, it's easy to forget that the weight of a nation's expectations are on these players' shoulders. It's only when you talk one-on-one with them that you realise how focused they are. Today, we're here to chat with three of that squad – Sol Campbell, Rio Ferdinand and Wes Brown. All three are used to dealing with the Premiership's top strikers, but now they will be expected to stop the world's best, if England are to progress in football's ultimate tournament.

The Introvert

Sol Campbell is the first to come over for a chat. Tall and imposing, he's remarkably quietly spoken, exuding the sort of assured calm in his speech that he displays on the pitch. Like

ROCK SOLID

most of Sven's squad, he refuses to talk up England's chances but admits to a certain optimism.

'There are a lot of quality, young players in this squad,' he says. 'Players who are world class or have the potential to become world class. No one's getting carried away but if we have a bit of luck, who knows?'

Sol talks with some authority, having already played a part in both the 1998 World Cup and Euro 2000.

'I learnt a lot from those tournaments, especially about the way I think during a game. When you face international teams you don't just compete on a physical level, you have to compete mentally as well. But the more you face that sort of challenge, the better you become as a player.'

And does Sol believe England can compete with supposedly more technically gifted sides?

He pauses before answering: '...If, say, an Italian defender hoofs a ball out of defence it's described as a tactical clearance. When an English player does the same, everyone claims it's because he's not technically good enough to pass his way out of defence. But what's the difference? Technically, the standard of English football isn't as bad as some critics make out. We're not far short of getting it right."

The Perfectionist

Few would claim Rio Ferdinand isn't technically gifted. The Leeds star is already acknowledged as one of Europe's finest defenders, but he isn't taking anything for granted.

'I've learned not to get carried away by comments, good or bad,' admits Rio. 'I was in the squad for France '98 but then left out for Euro 2000 and although things have been going okay recently, I won't get complacent. I want to be perfect and I hate making mistakes so I need to keep working and improving.'

'The move from West Ham to Leeds helped me, and particularly the chance to play in the Champions League. In Europe, or with England, the ball can be up the other end of the pitch for five minutes then – bang! – it's in the final third. Young players can switch off when the ball's not around them, but if you lose concentration for a split second in Europe it can cost you dear.'

Clearly, Rio is maturing into a fine international, but how has he enjoyed working under Sven?

'He's been brilliant for this squad,' enthuses Rio. 'Since Sven arrived I've seen a big change, with the younger players given more of a chance. He's even breathed new life into the fans. They expect a lot more from us now.'

The Future

Wes Brown is one of the many young faces regularly in Sven's England squad, although he's already packed more into his career than most players could ever hope to. He made his debut for Manchester United at the ripe old age of 19 at the tail end of the 1997–98 season.

'I'd always read the sports pages as a kid,' he grins. 'But it is weird when you start seeing your face in there. After a while though, you get used to it. Even if they put something stupid about you in the papers, I tend to laugh it off or just ignore it. It's just everyday life now, innit?'

In 1998–99, 'everyday life' included being part of United's Treble-winning squad. Unsurprisingly, an England call-up soon followed and, despite some injury knocks, Wes has returned stronger than ever. Many have even cited him as the future of Sven-Göran Eriksson's England defence.

Not that all the praise appears to have affected the laid-back Mancunian who seems almost bemused by the plaudits.

'It's nice, I guess, that people have said those things,' he smiles. 'But I'm still learning my trade. All I can do is train hard and hope the manager picks me.'

Wes admits however that the thought of going to Japan excites him.

'Oh yeah! It's what you dream about as a kid, going to play in the World Cup Finals. I don't think about it in great detail because I don't know for sure that I'll be picked but it would be great to go. I think we've got a team that can compete with the best. If the squad keeps progressing, we can win matches.'

And if Wes, Sol and Rio keep progressing, the future of England's defence looks in pretty good shape too.

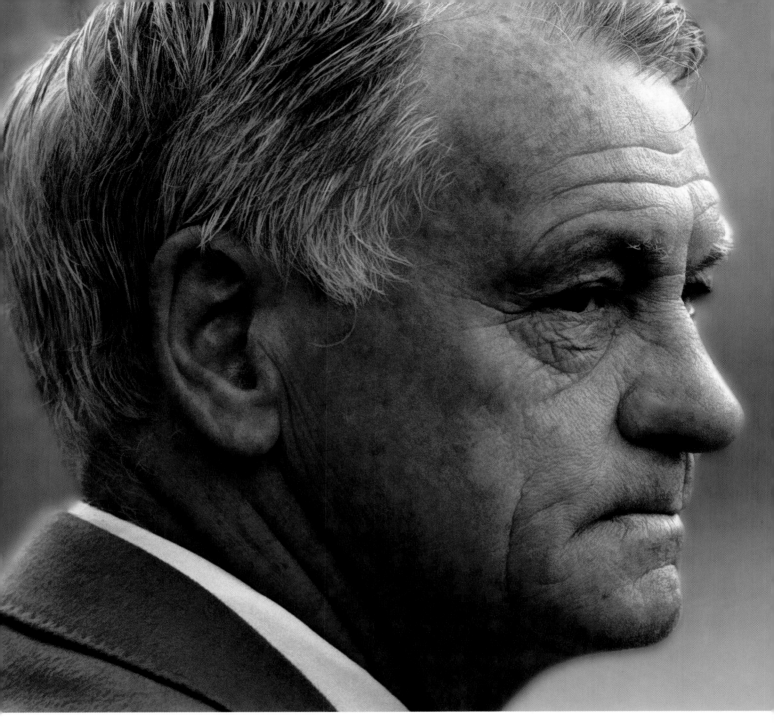

Been there, done that. The most successful living England manager, Bobby Robson, explains what it's like to lead a team to the World Cup, carrying the hopes of a nation with you...

Team spirit

'Team spirit is very important. I always said to my players, "Look, we're going away for six weeks. Anybody's who's going to be homesick, tell me now and don't come because we don't want people like that. We want people who know it's a six-week commitment and want to be there, want to get to the final, not get knocked out and go home early."

'It's very important you have a group that gets along well. A bit of friendship. But you don't start preparing for the finals in April: you work together for two years leading up to it. So the players, to some degree, are integrated already. They have a personal rapport. It should be fairly well bedded down before you go.'

Stay focused

'Boredom can be a factor at the team hotel. I've always found training very important in keeping players happy. Interesting training with the right balance of rest and relaxation and a certain amount of letting them do their own thing.

'And during a World Cup, there's always a game to watch! You know, you can train in the morning then go

BOBBY ROBSON

THE MAN WHO KNOWS

and watch France v Denmark in the afternoon.

'Your biggest problems are always with the players who aren't playing, not those who are picked. You've got to make them feel important, that they are needed and that they'll get their chance.'

Be the boss

'You are the coach, they are the players. There's always that line. You need a rapport with your players but there's only one guy who picks the team and decides the tactics. The squad has to know that. And to be honest, they like it. Players appreciate a coach who is assertive, and who can guide them.

'There were some suggestions of player power when I was with England, but there was no truth in that. We're intelligent people, I'd happily ask my players for suggestions, but in terms of selecting the side and deciding how they play, it was never in the hands of the players. If it were, you'd be dead. Might as well pack it in. "You want to be in charge? Okay, you're in charge. I'm off home."

'The coach is the only one pulling the strings. You are under more pressure than the players, both inside the camp and from outside elements. But at least you can step back and analyse. You're not playing the matches so you can work out things during the games. That's the deal: more time, but more pressure.'

Media pressure

'The media stuff was horrible at times. But I got through it and, to a certain extent, triumphed. To be honest, going through all that made me who I am. Dealing with that level of pressure made me capable of going to some of the biggest clubs in the world and succeeding. If I hadn't been through all that, I might not have been prepared for the likes of Barcelona.

'It's part and parcel of the job. The England manager is the Prime Minister of football. He's the guy who is in charge so there will always be critics. You've got to be strong, believe in yourself and have faith in what you are trying to do. I always believed that I would prove the doubters wrong.'

Back home

'When things are going well for you in the tournament, you know there is a positive feeling among the fans, but to be honest you're not really aware of what's going on back home. I remember in Italy 1990, Bryan Robson got injured and I sent him back to England. About two weeks later, I got a call from him. He said, "You won't believe what's going on here. The whole country's jumping up and down."

'When you are away trying to win a tournament, you don't normally know that. We were in a little hotel, concentrating, trying to get things right.'

Lasting the distance

'It's a relief to get through the group phase. But you have to try not to get complacent and remind yourself that you haven't achieved anything yet except to avoid complete disaster.

'Once you reach the knockout stage, every game is on a knife edge. You know that, with respect, the 'rubbish' is slowly being eliminated. There will be no more easy or comfortable games. But if I sense any fear of the opposition, I always say to the players, "What do they think of us?" You only ever say things to players that they want to hear. Never say anything that sows seeds of doubt in their mind.

'Of course, I just missed out on reaching the World Cup Final in 1990. There's a certain pride in getting so close but ultimately you can't help but think what might have been. I just hope Sven-Göran Eriksson and his team can go all the way this time.'

RESPECT DUE

RIVALDO — BRAZIL

Coaches may criticise Rivaldo for not tackling back, but fans simply love watching this genius show off his lavish array of skills. The 1999 World Footballer Of The Year's influence will be critical if Brazil are to overcome their qualifying traumas and challenge for a fifth World Cup victory.

TORD GRIP

RIGHT-HAND MAN

You've probably spotted indefatigable England coach Tord Grip at your team's stadium, so it's about time you were introduced to Sven's closest confidante...

WHEN SVEN-GÖRAN Eriksson collected his 2001 Swedish Soccer Personality Of The Year award, he thanked his England captain David Beckham for scoring *that* free-kick against Greece, but he reserved special praise for a less-celebrated 63-year-old colleague.

'Most importantly, I'd like to thank Tord Grip for all his help,' said Sven. 'He is a great guy.'

Tord Grip, Sven's right-hand man, prefers to keep a fairly low profile, but his contribution to England's World Cup qualification cannot be understated. His relationship with the England manager is so close that the Swedish media have nicknamed Tord, 'one half of Sven-Göran's brain' and Eriksson makes no attempt to hide Tord's influence on his work. When asked to nominate his own key quality as a coach, Eriksson replies simply, 'Tord Grip.'

And Tord's coaching record is impressive. The former Norway international coach has also had two spells as Sweden's assistant coach, working at the 1978 World Cup and returning to help his native country reach the semi-finals of the 1992 European Championship and USA 94.

'Tord is less well-known by the general public here in Sweden but he is well-known and respected by those in football,' says current Sweden coach Tommy Soderberg. 'Not only was he an influential part of the Swedish team that reached the 1994 World Cup semi-finals, but he has also coached in Italy, Switzerland and even Indonesia. Tord has a tremendous amount of experience and he is a calm and intelligent man.'

The former Swedish international player's working relationship with Sven-Göran Eriksson dates back to the 1970s when Tord was Player-Coach at KB Karlskoga and Eriksson played

THE BACKROOM BOYS

Coaching an international football team in the 21st century is a major operation and Sven-Göran Eriksson has plenty of helpers. Alongside Sven and Tord Grip on the coaching staff are Middlesbrough manager **Steve McClaren** and Liverpool's First Team coach **Sammy Lee**, who were both identified by The FA as being among the best up-and-coming coaches in England. Former Manchester United manager **Dave Sexton** has also joined the senior coaching set-up after a successful stint working with the Under-21s while Liverpool legend **Ray Clemence**, who kept 27 clean sheets in 61 internationals back in the 1970s and early '80s, is the squad's Goalkeeping Coach. The support team is made up of **Dr John Crane** (Consultant Team Physician), **Dr Tim Sonnex** (Team Doctor), **Alan Smith** and **Gary Lewin** (physios), **Steve Slattery** (masseur), **Martin Grogan** (kit manager) and **Michelle Farrer** (administrator).

behind him at right-back. Tord wasn't particularly impressed with Eriksson's ball skills though.

'He was not that talented a player when he started,' smiles Tord, 'but he was very ambitious. He trained by himself and got better and better until he got his chance and when he finally got into the team he stayed there. He had the same ambition as a coach. Sven-Göran was always very determined to succeed.'

Tord gave the future England manager his first coaching job as his assistant at Degerfors in 1976 and they have since worked together with Sweden Under-21s and at Lazio where they guided the Italians to a League and Cup double. But even when their careers have taken them in different directions, they have kept in touch developing a near telepathic understanding. 'I can read his mind and perhaps he can read mine,' says Tord.

So when Eriksson left Lazio to become England coach in January 2001, it was only natural he brought Tord with him. Tord's first duty in the England set-up was to help familiarise Eriksson with the English league scene, a task he approached with such energy that he attended 45 Premiership matches within the first three months. The number is nearer 200 now, all those motorway journeys and scouting reports have paid off with England's qualification for the World Cup finals and Tord can't wait for his Japanese summer.

'The draw couldn't be tougher, but in some ways that's good because we have to be on our toes from the start,' he says. 'I'm really looking forward to the games, and from what I've seen the players are just the same.'

So Tord, how does it feel to be a key man in the England set-up at an age when most blokes are contemplating retirement?

'I think Sven has the biggest job in football and I'm a small part of it so it's nice,' he demurs.

Small part? Tord's former protégé-turned-boss will definitely disagree.

ENGLAND WORLD CUP 2002 31

PHASE ONE

HERE WE GO!

We've been drawn in the toughest group of the lot, but Sven and the boys are ready to take 'em all on. Here's the lowdown on England's Group F opponents...

ARGENTINA

Argentina are number two in the FIFA world rankings and now they want to take France's World Cup crown...

ENGLAND FANS OF A NERVOUS disposition, please turn over the page now before we reel off the list of established Argentinian stars... Still with us? Okay don't say you weren't warned.

For a start you've got Batistuta (Roma) and Crespo (Lazio) up front. Then there's Veron (Manchester United), Gonzalez (Valencia), Gallardo (Monaco), Zanetti (Internazionale), Samuel (Roma), Sorin (Cruzeiro), Ortega (River Plate) and Ayala (Valencia) too.

Bit frightening, isn't it? Oh, and we nearly forgot to mention the young bucks who are also challenging for starting places in Marcelo Bielsa's side. Like the dazzling Boca Juniors play-maker Juan Roman Riquelme, the gifted Javier Saviola, who is currently strutting his stuff with Barcelona, and classy Villarreal striker Martin Palermo.

Of course, great individuals don't necessarily gel to make a great team. But judging by Argentina's imperious qualifying form, Bielsa's favoured 3–3–2–2 formation suits his superstars down to the ground. The 'Albiceleste' demolished all-comers in the supposedly tough South American section, winning 13, drawing four, losing just one game and finishing 12 points clear of their nearest rivals Ecuador. Their dominance and attacking style was further emphasised by a massive +27 goal difference. Colombia coach Francisco Maturana put it simply: 'It is hell playing against Argentina,' he despaired.

But are the hell-raisers ready for the so-called 'Group Of Death'? Technical Director José Peckerman, who has also coached Argentina to three consecutive World Youth Cup titles since 1995, thinks it will help the team to focus immediately.

'There are two possibilities,' he said after the draw in South Korea. 'You can get a draw which allows you to improve step by step and allows some mishap.

Or you can get a start like this one, with very demanding rivals, in which all are capable of reaching the second round. Then you have to be fully concentrated from the beginning, without giving any advantages.'

The President of the Argentinian FA, Julio Grondona, also expressed his belief that the tough draw will do his national team good:

'Before, they were coming to the World Cup finals as total favourites. Now it will be essential that Bielsa has all the players he wants, that they arrive in good form, without injuries.'

ARGENTINA ESSENTIALS

- Argentina have won the World Cup on two occasions (1978 and 1986) and finished runners-up twice (1930 and 1990).

THE ROAD TO WORLD CUP 2002

CONMEBOL, SOUTH AMERICAN QUALIFYING GROUP

29/03/00 v Chile (h) won 4–1
26/04/00 v Venezuela (a) won 4–0
04/06/00 v Bolivia (h) won 1–0
29/06/00 v Colombia (a) won 3–1
19/07/00 v Ecuador (h) won 2–0
26/07/00 v Brazil (a) lost 1–3
16/08/00 v Paraguay (h) drew 1–1
03/09/00 v Peru (a) won 2–1
08/10/00 v Uruguay (h) won 2–1
15/11/00 v Chile (a) won 2–0
28/03/01 v Venezuela (h) won 5–0
25/04/01 v Bolivia (a) drew 3–3
03/06/01 v Colombia (h) won 3–0
15/08/01 v Ecuador (a) won 2–0
05/09/01 v Brazil (h) won 2–1
07/10/01 v Paraguay (a) drew 2–2
08/11/01 v Peru (h) won 2–0
14/11/01 v Uruguay (a) drew 1–1

TOP SCORER:
Hernan Crespo, 9 goals

FINAL POSITION:

P	W	D	L	F	A	GD	Pts	Psn
18	13	4	1	42	15	+27	43	1st

GROUP F FIXTURES
2 June v Nigeria, Ibaraki, 10.30am
7 June v England, Sapporo, 12.30pm
12 June v Sweden, Miyagi, 7.30am

MANAGER
MARCELO BIELSA
With the depth of quality in his squad, it is arguable that Bielsa has the easiest job of any coach at this year's World Cup. But he will be under serious pressure, because Argentina fans expect nothing less than a third World Cup triumph.

ONE TO WATCH
JUAN VERON
Manchester United's record £28.1 million signing has been hailed by George Best as the best player in the world. The midfielder's athleticism, silky passing skills and venomous shooting ability make him one of the key players in Argentina's brilliant line-up.

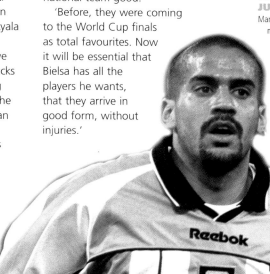

Beware the silky skills of 'Seba' Veron.

PHASE ONE

🇳🇬 NIGERIA

In 1994 and 1998, the Super Eagles reached the second round of the tournament. Can they do even better this time?

MANAGER
SHAIBU AMODU
Amodu helped Nigeria to recover from a terrible start to their World Cup 2002 qualifying campaign as temporary coach and has been promised a permanent position if the Super Eagles perform well in Korea and Japan.

ONE TO WATCH
JAY-JAY OKOCHA
The gifted 28-year-old midfielder's dribbling skills delight fans and drive opposing defenders mad. 'Jay-Jay' has spent the past decade playing his club football in Germany, Turkey (where he won the 1996–97 league title with Fenerbahçe) and most recently for Paris St Germain in France.

Gifted midfielder Jay-Jay Okocha can drive opposing defenders mad.

NIGERIA ESSENTIALS

◎ In the '90s, Nigeria won gold at 1996 Olympics and twice won the African Nations Cup in 1994 and 1998.

◎ Nigeria first qualified for the World Cup finals in 1994 and this will be their third consecutive appearance.

THE ROAD TO WORLD CUP 2002

CAF AFRICAN QUALIFYING GROUP 2

17/06/00	v S Leone (h)	won 2–0
09/07/00	v Liberia (a)	lost 1–2
27/01/01	v Sudan (h)	won 3–0
09/03/01	v Ghana (a)	drew 0–0
21/04/01	v S Leone (a)	lost 0–1
05/05/01	v Liberia (h)	won 2–0
01/07/01	v Sudan (a)	won 4–0
20/07/01	v Ghana (h)	won 3–0

TOP SCORER
Victor Agali, 5 goals

FINAL POSITION

P	W	D	L	F	A	GD	Pts	Psn
8	5	1	2	15	3	+12	16	1st

GROUP F FIXTURES
2 June v Argentina, Ibaraki, 10.30am
7 June v Sweden, Kobe, 7.30am
12 June v England, Osaka, 7.30am

SINCE THEY FIRST QUALIFIED FOR the finals in 1994, Africa's most populous country has been regarded as the continent's best prospect of World Cup success. But Nigeria will certainly need to raise their game to have a chance of progressing in Korea/Japan after an unconvincing and controversial qualifying campaign.

They were up against Liberia, Sudan, Ghana and Sierra Leone in Group Two of the CAF section and started as hot favourites. But two defeats in their first five matches, including an astonishing 0–1 away loss to group whipping boys Sierra Leone put them on the back foot. They showed fighting spirit to produce three wins in their last three fixtures to top the group though. However, this achievement was overshadowed by allegations that Ghanaian officials and players received a $25,000 cash 'gift' from a Nigerian state governor after Nigeria's last-match win over Ghana.

But Nigeria have qualified for their third consecutive World Cup and if they get their act together they are capable of causing an upset or two. Most of the Nigerian internationals play their club football abroad so there is a breadth of experience in the squad. And the team should also show the benefits of putting their faith in youth when Nigeria made their first appearance in the World Cup finals back in 1994. The likes of Sunday Oliseh, Jay-Jay Okocha, Finidi George and Victor Ikpeba were all raw talents back then, but with another World Cup and an triumphant Olympic campaign (1996) under their belts, they know all about handling the pressure of a big tournament. And when you throw in the creativity of Nwankwo Kanu plus the finishing of Victor Agali, Nigeria can be a match for anyone.

Their fans are optimistic too. Rauf Ladipo, leader of the Nigerian Football Supporters Club has no fears about Group F. 'There is no big deal about the teams,' he says. 'I know we will qualify for the second round!'

SWEDEN

Our Sven's home country were unbeaten in qualifying for World Cup 2002 and will offer stern resistance in England's opening match...

Celtic star Henrik Larsson (left) adds a cutting edge to a solid Sweden team.

'WELL, WHAT CAN YOU SAY... It will be fun for all football enthusiasts!' That was the upbeat verdict of Sweden co-coach Tommy Soderberg on being drawn in the 'Group Of Death'. And he has every reason to feel optimistic, because Sweden were one of the most impressive qualifiers for this year's tournament. Even allowing for the fact that they were drawn in one of the weaker UEFA groups, a record of eight wins and two draws in 10 matches was an exceptional effort.

The team's success is based on team play, defensive organisation and their under-rated goalkeeper Magnus Hedman. Hedman has finished on the losing side just four times in 41 internationals and he conceded just three goals in the qualifiers. In front of him, captain Patrik Andersson marshalls the defence with cool assurance.

Twenty goals in 10 qualifiers shows they carry an offensive threat too. Freddie Ljungberg provides energy and invention in midfield, while up front, ace marksman Henrik Larsson has formed a potent partnership with Heerenveen's Marcus Allbäck. Also, look out for Ajax's 19-year-old striking prodigy Zlatan Ibrahimovic.

Critics in the English media have been guilty of writing off Sweden's chances pointing to their failure at Euro 2000. But such a judgement ignores the fact that Sweden, who conceded just one goal in qualifying, went into that tournament with key players like Larsson and Ljungberg barely fit. Such complacent thinking also conveniently washes over England's failure to beat Sweden in nine attempts since 1968.

'In the English papers I've read pundits saying "Sweden is a small country, we should beat them",' notes Magnus Hedman. 'I like that. If England want to think they are better, that suits us.'

If we do knock out the Swedes though, we can count on the support of some of their fans. Before our 1–1 friendly draw last year, a Swedish newspaper did a poll asking readers who they would be supporting? Some 26,000 replied and a fifth said they were backing England!

MANAGER
TOMMY SÖDERBERG & LARS LAGERBÄCK

Sweden are the only team with joint coaches, but it seems to work due to their laid-back temperaments. When Soderberg was asked who had been the most important person in Sweden's qualifying campaign, he replied, "The woman in our office who books all our trips!"

ONE TO WATCH
HENRIK LARSSON

The 30-year-old Celtic star (pictured) has proved himself to be one of Europe's most dangerous strikers at club and international level. Larsson's goal threat added to his team's defensive solidity make Sweden formidable opponents.

SWEDEN ESSENTIALS

◎ Sweden's best-ever World Cup performance was to reach the final of the 1958 World Cup staged in Sweden.

◎ Goalkeeper Magnus Hedman has the amazing record of keeping 32 clean sheets in 41 internationals.

THE ROAD TO WORLD CUP 2002

UEFA QUALIFYING GROUP 4

02/09/00 v Azerbaijan (a)	won 1–0
07/10/00 v Turkey (h)	drew 1–1
11/10/00 v Slovakia (a)	drew 0–0
24/03/01 v Mac'donia (h)	won 1–0
28/03/01 v Moldova (a)	won 2–0
02/06/01 v Slovakia (h)	won 2–0
06/06/01 v Moldova (h)	won 6–0
01/09/01 v Mac'donia (a)	won 2–1
05/09/01 v Turkey (a)	won 2–1
07/10/01 v Azerbaijan (h)	won 3–0

TOP SCORER
Henrik Larsson, 8 goals

FINAL POSITION

P	W	D	L	F	A	GD	Pts	Psn
10	8	2	0	20	3	+17	26	1st

GROUP F FIXTURES

2 June v England, Saitama,	10.30am
7 June v Nigeria, Kobe,	7.30am
12 June v Argentina, Miyagi,	7.30am

STEVEN GERRARD

LIONHEARTS

THE ENGINE ROOM

The emergence of young STEVEN GERRARD as England's premier ball-winning midfielder was a key factor in England's qualification for World Cup 2002. Veron, Ljungberg and co better watch out...

'Argentina are good but we are capable of beating any team on our day'

WHEN YOU MEET FOOTBALLERS, they are often very different physically from how you'd expect. Usually they are shorter (just like film stars) and less bulky (even Gazza) due to the strict training and dietary requirements of the modern game. In person, Steven Gerrard clearly exudes the good health of an athlete with an unfeasibly low body fat ratio, but despite his tendency to hunch his shoulders, his slim six-foot frame makes him as conspicuous a figure off the pitch as he is in England's midfield.

And during our chat in the lobby of the Burnham Beeches hotel, England's training base before internationals, he radiates the quiet confidence of a young man who has become a key figure in Sven-Göran Eriksson's plans for World Cup 2002. Steven will celebrate his 22nd birthday three days before our opening fixture against Sweden and appears unfazed by the tough opposition England will face in Group F.

'Before it even happened, everyone could see how it was going to turn out,' he smiles remembering the day of the draw last December. 'There were just three or four balls left and there was just this feeling for everyone in the room that we were going to draw Argentina.

'Obviously we'd have preferred Japan or Korea's groups, but it was not to be. There are four fantastic teams in there, but if we want to be the best we have to beat the best. We're just going to have to do it sooner than expected. Argentina are good but we know that we are capable of beating any team on our day. It's vital we get off to a good start against Sweden in our first game and we are looking forward to the challenge.'

No fear at all then, but you wouldn't expect any from a player who relishes the heat of a midfield battle. Like many of his peers, Steven's strongest World Cup memory is of Paul Gascoigne's wizardry at Italia '90, but he grew up knowing his style of play was more in keeping with boyhood hero Steve McMahon. And since his emergence with Liverpool and England, his combative style has already drawn comparisons with Roy Keane and Patrick Vieira.

'I admire Keane and Vieira most, because they play the way I've always wanted to play,' he says. 'And I enjoy playing against players like them, contesting the same area of the pitch.'

Like his two midfield rivals, aggression and competitiveness are important parts of Steven's game, but he is fully aware of the need for self-control.

'There have been times when I've lost my head and risked getting sent off, which was bad for the team,' he admits. 'Maybe it was down to over-enthusiasm in certain games, trying hard to impress. But I don't get told to take the aggression out of my game, just to calm it down.'

Anyway, like Keane and Vieira, Gerrard has far more footballing assets than just steel and passion as Tord Grip notes.

'Steven is a good athlete, good in the air, a good passer of the ball and he scores goals too,' says Sven's right-hand man.

Hmm, sounds suspiciously like the complete midfielder, but Steven is wary of getting carried away by the praise that has been heaped upon him. Indeed, he is the first to admit that following his stunning rise to prominence during Liverpool's Treble-winning season his form was up and down in the early part of 2001–02.

'I've been hot and cold this season, not as consistent as I would like to be,' he ponders. 'Maybe I was just trying too hard in certain games. But I have set myself high standards, and no one keeps going up and up. That's the way I want it to be though. I'm not complaining about expectations being high, I'm happy about it.'

And it's not only the public and media who expect a lot of him, there is also England manager Sven-Göran Eriksson. But according to Steven, with Eriksson there's no pressure, it's all about belief:

'He believes in this squad and all the players believe in him too. He is always so calm and confident.'

No wonder, when he has players of Steven Gerrard's calibre at his disposal.

ENDURING IMAGE

Was it or wasn't it?

30 July 1966, Wembley
World Cup Final
England 4 West Germany 2 (AET)

Geoff Hurst's 100th-minute shot hits the crossbar and the ball bounces down over the goal-line (well, the Russian linesman thought so anyway!) to give England a 3–2 lead on our way to a famous World Cup triumph.

THE PHYSIO

HANDLE WITH CARE

The England medical team will be looking after millions of pounds' worth of footballing talent at World Cup 2002. Physio GARY LEWIN explains how they give players plenty of TLC...

I'M SITTING WITH ENGLAND physio Gary Lewin in the Massage Room at Burnham Beeches hotel. Well, I say 'Massage Room' – in fact, it's usually just a big, empty suite. But whenever the England team come to train at the hotel, Gary and the rest of the medical team convert it into a treatment room and the room next door into a medical facility.

There is activity all around us. On the two beds, masseurs Steve Slattery and Rod Thornley are kneading the legs of Messrs Scholes and Gerrard. To their right, captain David Beckham is talking ligaments with one of the England squad's two doctors Tim Sonnex (the other, Dr John Crane has popped off to give Sven an injury report). Through the door, England's other physio Alan Smith is pulsing electrical currents through Sol Campbell's legs. Clearly, keeping the England players fit is a real team effort.

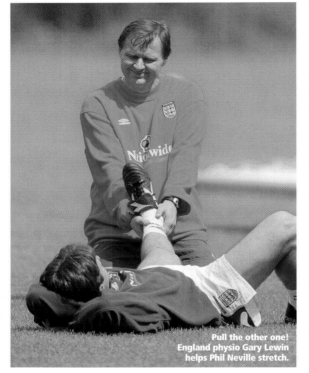

Pull the other one! England physio Gary Lewin helps Phil Neville stretch.

'Tim, John, Steve, Rod, Alan and myself work very closely together,' agrees Gary. 'And we also liaise closely with physios at the players' clubs. They will call Alan and me the day before their players report to let us know any ailments – we don't want to provide conflicting treatment without the club's consent. Then after the match we'll advise them if a player has picked up any knocks.'

A former apprentice goalkeeper at Arsenal, Gary opted for a career in physiotherapy when he was released at the age of 18. After 'playing for Barnet and scraping through two A levels in a year', he started a three-year course at Guy's Hospital Of Physiotherapy, qualifying in June 1986, and returned to Arsenal as a physio that September. A decade later, Glenn Hoddle offered him the chance to work for England alongside his friend, Sheffield Wednesday physio Alan Smith.

Gary says his proudest England moment so far was the 5–1 win over Germany last year. And he really made a big contribution to that victory by helping the England skipper David Beckham get fit to train and then play.

'David sustained a knee injury on the Sunday before Saturday's game,' recalls Gary before explaining the rigorous checks applied to Becks and all injured England players. 'He had treatment on Monday, Tuesday and Wednesday, but each day we reassessed him and deemed him unable to take a fitness test.

'On Thursday, we went out for a light jog and then travelled to Germany. On Friday morning, we felt he was able to take a fitness test and David passed. That still didn't mean he was fit for selection, but he could train with the team that afternoon. Happily, he did so with no reaction and the doctors advised Sven he could play on Saturday.' Phew!

Gary and co are now making meticulous preparations for World Cup 2002. Just as at Burnham Beeches, they will set up their own 'medical centre' at England's pre-tournament hotel in Dubai and then at their base in Japan for the tournament proper.

'It's like M.A.S.H. – we just take over a room!' laughs Gary. 'At France '98 we had a nice villa to ourselves though. The F.A. have done thorough recces out in Japan and they will make sure we have a gym and swimming pool facility etc. And we have to make sure we take all the physiotherapy and First Aid equipment we need.'

Because players spend a lot of time hanging around the physio room, the medical staff will also fulfil a valuable role in helping England boss Sven assess team morale during the tournament.

'An experienced physio can tell if a player is having problems,' he says. 'It's easier to spot at club level when you are with players all season, but when you go away to a World Cup, you become a little football club for a few weeks and you want to help all the players fit in.'

RESPECT DUE

GABRIEL BATISTUTA — ARGENTINA

'Batigol' has been a consistent scorer at the highest level for more than a decade and is accorded God-like status in Argentina. He fired in five goals at France '98 and the 33-year-old will form a formidable partnership with Hernan Crespo at this year's tournament.

PARTY TIPS

LET'S PARTY!

Not going to Japan for the finals? Worried about creating a match atmosphere in the early hours of the morning? Have no fear, our opposition-themed party tips could be the answer (then again, they might not help at all...)

ENGLAND v SWEDEN
JUNE 2, 10.30 AM

MUSIC
Okay this first tip is obvious but still relevant: Benny, Bjorn, Agnetha and Frida. That's right, stick a bit of Abba on the stereo and within the first few bars of *Waterloo*, your guests will have completely forgotten that it's only the middle of the morning (er, probably). If that seems a bit too obvious, try a more obscure Swedish band. We'd recommend the rather excellent Whale – their album *We Care* should do the trick.

FOOD
Ideally, you'll get the Swedish chef from *The Muppet Show* round to do the

cooking, but if for any reason he proves to be unavailable, don't panic. It's relatively easy to put together a smorgasbord for your guests. This consists of pickled Baltic herring as a centrepiece, surrounded by seafood, salads, cold cuts, boiled eggs and cheeses. Acceptable additions are meatballs, and a combination of anchovies and scalloped potatoes known as Jansson's Frestelse, or Jansson's Temptation. Oh, crispbread's good as well.

DRINK
The Swedes like their alcohol and Sweden is home of trendy Absolut vodka. Of course, you could always fill one of those beautifully designed bottles with water then swig away to your heart's content while amazing friends and colleagues with your apparent ability to gulp down the hard stuff at breakfast time...

CLOTHING
To get that authentic Swedish look for your party, we suggest you get hold of some Bjorn Borg underwear. Invite all your best looking friends round and suggest that they model it for you. How, you ask? Um, we're still trying to work that one out ourselves... Alternatively, get your nan to embroider you a folksy looking jerkin (okay, we know that's a rather poor second, but it's probably a wee bit easier to arrange).

AND ANOTHER THING...
If the unthinkable happens and England don't get the desired result, don't try and shake that melancholy feeling. Wallow in misery with the help of Sweden's most famous film director, Ingmar Bergman. The master of depressing introspection has a range of films that will keep you feeling low all day. Well, it's either that or go into work.

RECIPE
Janssons Frestelse
INGREDIENTS
6 to 8 potatoes, butter, 2 onions, 2 cans of anchovy fillets, 1 to 1 1/2 cups of light cream

Instructions
Peel the potatoes and cut into thin sticks. Slice the onions and sauté in some of the butter. Drain the anchovies and cut into pieces. Put the potatoes, onion and anchovies in layers in a buttered baking dish. (The first and last layer should be potatoes.) Dot with butter on top. Pour in a little of the juice from the anchovies and half of the cream. Bake in a pre-heated oven at 200°C for 20 minutes. Pour in the remainder of the cream and bake for another half hour or until the potatoes are tender. Serve as a first course.

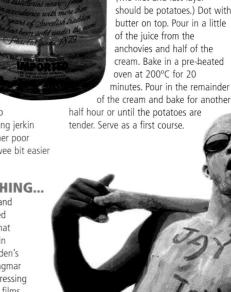

ENGLAND v ARGENTINA
JUNE 7, 12.30PM

MUSIC
Argentina is the home of the tango, a dance and music popular with romantics around the world. So pick up a suitable recording and proceed to dance cheek to cheek in a melodramatic fashion for ten minutes, or until tired/embarrassed/emotional. If you're watching the game down the pub with your mates, you may want to skip this and just discuss Owen's wonder goal in '98, Becks' sending off etc.

FOOD
A lot of Argentine dishes reflect the Italian heritage of much of the country. Pasta, ricotta, and polenta are all popular. Veal is also a regular, as in Carbonada Criolla (Veal and Vegetable Stew). For hors d'oeuvres, try a picada – small plates with simple, easy-to-get-hold-of fare such as French bread, potato chips, pâté, cheese triangles, peanuts, ham, and pickled onions (yes, pickled onions). Another touch of authenticity can be obtained with Dulche de Leche (Milk Jam). Simply get a tin of sweetened condensed milk, put in a pan of boiling water, simmer for four hours and, er, Roberto's your uncle. (Make sure you keep the tin immersed in water though or it'll explode.) If all that sounds too complicated, have a corned beef sandwich instead.

DRINK
Argentina produces some very fine wines, and if you pop down to your local offy you'll find a number of decent reds for under a tenner. We recommend a fruity Malbec and 2000 was a vintage year.

CLOTHING
Um, we don't know much about Argentine clothing so we checked out one of the country's trendiest boutiques. Apparently, there are four styles that a dedicated follower of Argentine fashion would currently favour: 'Military' (fairly self-explanatory), 'Black and White' (likewise), 'Flowery' ('like hippy sixties', they said) and 'Neopunk' ('think London chick' were the words of wisdom on that one...). None the wiser? Neither are we. Just don't walk down the street in blue and white stripes.

AND ANOTHER THING...
Again, we're sure England will win handsomely, but if a brave face is needed in the face of adversity, put your hand on your heart and begin singing a blubbery version of *Don't Cry For Me Argentina*. After all, Japan is the spiritual home of karaoke.

RECIPE
Carbonada Criolla
INGREDIENTS: 2 medium onions, chopped, 3 cloves of garlic, mushed, 1 kilogram of veal, 1 can of whole peeled tomatoes, chopped, salt and pepper, 50 grams of butter, oil, 2 sliced carrots, 1 sweet potato, cut up, 1/4 kg of squash (pumpkin-type thing), cut up, 3 potatoes, cut up, 2 corn ears, cut up, 1 cup of rice, 1 can of peaches, beef broth.

Instructions
Brown the onion and garlic in the oil and butter. Cut the meat into squares and add to the pan. Brown well. Incorporate the tomatoes, seasonings, carrots and beef broth to cover the ingredients. Cook for 15 minutes and add the rest of the vegetables. Cook until the vegetables are half done then add the rice and the peaches. Continue simmering until everything is perfectly cooked.

ENGLAND v NIGERIA
JUNE 12, 7.30AM

MUSIC
Without wanting to get all Andy Kershaw, Nigeria has produced some pretty stonking music, a lot of which is quite easy to get hold of should you wish to create the right ambience for England's third match of the tournament. Fela Kuti's fusion of freestyle jazz and Yoruba call and response chanting should do it. Or try a bit of juju music, courtesy of Sonny Ade. If that all sounds a little elaborate, just dig out one of your old Sade albums.

FOOD
Impress friends and influence neighbours by picking up a bottle of Zulu Fire Sauce (yes, really). You can get it from some supermarkets and it looks the part, with its scary mask on the label and its ethnic beads and tribal neckwear below the bottletop. Not sure how true it is to Nigeria though. For the real deal try preparing some Isi-ewu (Goat Head Pepper Soup). Beef, maize, grains, corn rice and millet are all Nigerian staples too while bananas, pineapples, tangerines, guava, melons, grapefruits and mangoes are also popular.

DRINK
Palm wine, a drink made from the juices of palm trees is very common in Nigeria although possibly a little tricky to obtain in Blighty. You could get away with pineapple juice or orange juice (which, let's face it, is probably a bit more palatable at breakfast time).

CLOTHING
A tricky one this. We suggest you simply turn up the central heating and don't wear too much. Perhaps a Yoruba mask if you can get hold of one. The reality is most Nigerians wear Westernised clothing anyway. You could try adding a lucky charm such as the ibej (twin dolls that are adorned with beads), if you're really after that authentic thang...

AND ANOTHER THING...
Guinness is very popular in Nigeria, although this 'foreign stout' is much stronger out there – same brand and all but twice as strong as the draught (which they don't have). This, combined with the temperature, is responsible for the hospitalisation of several western travellers every year.

RECIPE
Chin Chin and Puff Puff
INGREDIENTS
Flour, sugar, butter, egg yolk

Instructions
Puff Puff: Mix the flour, sugar, butter and egg yolk in a bowl until you achieve crumb-like consistency. Use your hands to bring together the mixture and knead as a dough. Leave to rest for a while in the fridge then separate into eyeball-sized portions. Deep fry until golden. Drain and rest on some kitchen roll. Dredge in icing sugar.
Chin Chin: Achieve the same dough but replace the sugar with a teaspoon of salt. Roll out the dough then cut up into small bitesize squares. Shallow fry in oil until a deep golden brown. We recommend dipping these in Zulu Fire Sauce to add a fiery tang.

ENGLAND WORLD CUP 2002 43

BIG GUNS

THE CON

The World Cup brings together the élite of international football.

BRAZIL

After a surprisingly traumatic qualifying phase, Brazil finally made it to the World Cup and history suggests they'll perform rather better in the tournament itself.

MANAGER
FELIPE SCOLARI
The third coach to be employed by Brazil during the qualifying campaign, Scolari just managed to get his under-achieving superstars through to the finals. Now, can he find a system to exploit fully the attacking talent at his disposal?

ONE TO WATCH
RONALDO
Since France '98, Ronaldo has suffered a string of injuries. While opposing defenders may disagree, all true football fans will be hoping to see the former European (1997) and World Footballer Of The Year (1996 & 1997) back to his brilliant best.

FOR A WHILE IT LOOKED LIKE Brazil might not make it to Japan and Korea. Never in the country's history had they come as close to being eliminated before the finals, with an astonishing six defeats. Fortunately for lovers of their unique style of football, a 3–0 victory in the decisive match against Venezuela ensured the Brazilians wouldn't be on holiday this summer. The result was all that mattered on the day but since then, the post-mortem on their inconsistent performances has begun.

Changes of coach didn't help: Wanderley Luxemburgo, Leao and then Felipe Scolari took charge in quick succession. But more detrimental has been the crisis of confidence affecting football in Brazil. For so many years, their fluent, attacking style didn't concern itself with such mundanities as tactical football. But is that a realistic option in the modern era?

It's a question Brazil will have to answer in Japan and Korea. What isn't in doubt however is the quality of the players they can call on: Barcelona's Rivaldo, Real Madrid's Roberto Carlos, Vasco De Gama's Romario, Roma's Emerson and Cafu, Flamengo's Edilson, Real Betis's Denilson… the list is long and intimidating. And then there's Ronaldo – will this be the tournament where the Brazilian striker fulfils his immense potential after a frustrating series of injuries?

Although coach Felipe Scolari may have had to put up with criticism in qualifying, all will be forgiven if Brazil return home with the World Cup. And who would dare write them off? The experience and talent is there and a good start to the tournament could see the players' confidence flood back. Reports of their demise may just have been exaggerated.

Romario (left) and Ronaldo… two fairly decent strikers.

BRAZIL ESSENTIALS

THE ROAD TO WORLD CUP 2002

CONMEBOL SOUTH AMERICA QUALIFYING GROUP

28/03/00 v Colombia (a) drew 0–0
26/04/00 v Ecuador (h) won 3–2
04/06/00 v Peru (a) won 1–0
28/06/00 v Uruguay (h) drew 1–1
18/07/00 v Paraguay (a) lost 1–2
26/07/00 v Argentina (h) won 3–1
15/08/00 v Chile (a) lost 0–3
03/09/00 v Bolivia (h) won 5–0
07/10/00 v Venezuela (a) won 6–0
15/11/00 v Colombia (h) won 1–0
28/03/01 v Ecuador (a) lost 0–1
25/04/01 v Peru (h) drew 1–1
01/07/01 v Uruguay (a) lost 0–1
15/08/01 v Paraguay (h) won 2–0
05/09/01 v Argentina (a) lost 1–2
07/10/01 v Chile (h) won 2–0
07/11/01 v Bolivia (a) lost 1–3
14/11/01 v Venezuela (h) won 3–0

TOP SCORERS
Rivaldo / Romario, 8 goals each

FINAL POSITION

P	W	D	L	F	A	GD	Pts	Psn
18	9	3	6	31	17	+14	30	3rd

GROUP C FIXTURES
3 June v Turkey, Ulsan, 10am
8 June v China, Seogwipo, 12.30pm
13 June v Costa Rica, Suwon, 7.30am

TENDERS

Let's take a closer look at the big guns, dark horses and the rest...

FRANCE

The current European and World champions have already proved their ability but can they maintain their success?

FRANCE ESSENTIALS

◎ Before 1998, France had never won the World Cup. Their best placing was third, in both 1958 (Sweden) and 1986 (Mexico).

◎ Four Frenchmen have won the European Footballer Of The Year award: Raymond Kopa (1958), Michel Platini (1983/84/85), Jean-Pierre Papin (1991) and Zinedine Zidane (1998).

◎ Euro 2000 was France's second triumph in the European Championship. They had previously won the tournament in 1984. In 1984, France also won the gold medal at the Olympics.

THE ROAD TO WORLD CUP 2002

◎ Not applicable – France qualified automatically as the reigning World Champions.

GROUP A FIXTURES
31 May v Senegal, Seoul, 12.30pm
6 June v Uruguay, Busan, 7.30am
11 June v Denmark, Incheon, 7.30am

AS BOTH EUROPEAN AND World champions, France remain the team to beat. But is their star on the wane? Coach Roger Lemerre will be hoping his team can provide a resounding 'no' to that particular question.

He certainly has the players to do that. Up front, Arsenal's lethal Thierry Henry has already proven his ability to convert the slimmest of chances. Under Arsène Wenger he has developed into the complete striker, while his colleague Patrick Vieira is arguably the finest midfielder in the Premiership. Combative and skilful in equal measure, Vieira could well use this tournament to cement his reputation worldwide. With Real Madrid's Zinedine Zidane in the centre too, few teams can claim a better central midfield partnership.

As Euro 2000 proved, the French are a potent force when they break forward but they'll need to recapture the defensive solidity they displayed in 1998 if they are to hold on to their World Cup. Laurent Blanc's retirement from international football has allowed his Manchester United team-mate Mikael Silvestre to earn a regular first-team spot. Silvestre has come in for some criticism for his defensive work but regular watchers of the Frenchman will know that he is tremendous going forward. Alongside him, Marcel Desailly is still more than capable and with Bixente Lizarazu and Lilian Thuram as full backs it's unlikely that too many strikers will unlock the French defence.

Even if they do, there's still the imposing, inspiring Fabien Barthez to beat. Despite his unpredictability, the fact remains that there are few better shot-stoppers in the world. If he displays his usual gymnastics in goal, the French team will again be very hard to overcome.

MANAGER
ROGER LEMERRE
A French international player in the '70s, Roger Lemerre had the unenviable task of succeeding his guru Aimé Jacquet after France '98. But he simply continued Jacquet's good work, guiding France to another famous triumph at Euro 2000.

ONE TO WATCH
LILIAN THURAM
Defenders don't normally attract a lot of attention, but 30-year-old Lilian Thuram is no ordinary player. Athletic, aggressive and skilful, Thuram can play in any position across the defence. That's why Juve paid a whopping £22 million for his services last summer.

Golden memories but can France retain the World Cup?

BIG GUNS

ITALY

With their vice-like defence, a number of gifted strikers and the visionary influence of Francesco Totti, are Italy capable of winning their fourth World Cup?

Class personified: the legendary Paolo Maldini

MANAGER
GIOVANNI TRAPATTONI

One of the most successful Italian coaches of all time at club level, Trappatoni will be hoping to erase the memory of Italy's painful extra-time defeat by France in the Euro 2000 final and he certainly has the tactical nous to ensure the 'Azzurri' are as hard to beat as ever.

ONE TO WATCH
PAOLO MALDINI

One of the world's greatest defenders, Paolo Maldini is practically footballing royalty – his father, Cesare Maldini, was a star for AC Milan and Italy in the 60s before becoming national coach. Maldini Jnr has won five Italian championship medals among a whole host of honours, and his perfectly-timed tackles and reading of the game are textbook stuff.

ITALY ESSENTIALS

◎ Italy have reached the World Cup Final five times and won the tournament three times – in 1934, 1938 and 1982.

THE ROAD TO WORLD CUP 2002

UEFA QUALIFYING GROUP 8
03/09/00 v Hungary (a) drew 2–2
07/10/00 v Romania (h) won 3–0
11/10/00 v Georgia (h) won 2–0
24/03/01 v Romania (a) won 2–0
28/03/01 v Lithuania (h) won 4–0
02/06/01 v Georgia (a) won 2–1
01/09/01 v Lithuania (a) drew 0–0
06/10/01 v Hungary (h) won 1–0

TOP SCORER
Alessandro Del Piero, 5 goals

FINAL POSITION
P	W	D	L	F	A	GD	Pts	Psn
8	6	2	0	16	3	+13	20	1st

GROUP G FIXTURES
3 June v Ecuador, Sapporo, 12.30pm
8 June v Croatia, Ibaraki, 10am
13 June v Mexico, Oita, 12.30pm

LIKE BRAZIL, ITALY ENDURED less-than-convincing performances in qualifying. But, like Brazil, history suggests that they will perform better in the tournament proper. Certainly, among Italian fans, confidence is high that their side are real contenders.

Despite the tentative qualification, the fact remains that this Italian side is very similar in personnel to the team that reached the final of Euro 2000 and technical director Giovanni Trapattoni has an embarrassment of riches from which to pick his team.

Tight defence will, as usual, be the backbone of the Italian's tactics – it's hard to lose matches when you so rarely concede goals (in Euro 2000, they let in just three in eight matches).

A lot of the credit for such a tight rearguard goes to Paolo Maldini, Italy's most capped player, who will retire from international football after this tournament. The 33-year-old has been the most consistent Italian footballer of the last decade. His natural ability, pace and reading of the game have helped to secure many a clean sheet.

Keeping Maldini company at the back are likely to be Fabio Cannavaro and Alessandro Nesta. A challenge for even the most potent of frontmen.

In front of this stronghold will be Francesco Totti, hailed by coach Trapattoni as a 'visionary leader, a player capable of inspiring his side to World Cup victory'. He certainly was an inspirational figure for Roma during the club's Scudetto-winning campaign with his ability to lift the players around him, while his passing and commitment are exemplary.

Up front, Christian Vieri has been Trapattoni's preferred striker, the only cloud on the horizon being a proneness to injury. Therefore, there might be opportunities for some other familiar faces, such as Alessandro Del Piero and Filippo Inzaghi and even the old master Roberto Baggio. One thing is sure – Trapattoni won't be short of talented players to choose from.

SPAIN

With their domestic league in such good shape, will this finally be the time for perennial underachievers Spain to come good at a World Cup?

The lad Raul knows where the goal is...

IT'S BECOME A WORLD CUP tradition to talk up Spain's chances of success only for the team to disappoint. Whether this time will be any different is anyone's guess but the signs are promising for the Spaniards. Their domestic league is arguably the best in Europe, and in Raul they have one of the finest players in the world.

To date, Spain's only international successes were the European Championship triumph in 1964 and Olympic gold in 1992. Both victories came on home soil. They are yet to reach a World Cup final.

But, despite falling to France 2–1 in the quarter-finals of Euro 2000, they showed enough promise to suggest they were only a couple of players away from being the finished article.

Those players have now arrived with the emergence of Deportivo La Coruña's Juan Carlos Valeron, Real Madrid's Ivan Helguera and Valencia's Vicente Rodriguez. Together with the established Raul, Fernando Hierro and Josep Guardiola, coach Jose Antonio Camacho has certainly got the players and an attacking system to suit them.

Two Italian-based players, Pep Guardiola and Gaizka Mendieta, could also figure in Camacho's plans. Since leaving Barcelona for Brescia, Guardiola has been impressive and although Mendieta struggled initially following his big-money transfer to Lazio last summer his ability isn't in doubt.

And coach Camacho is adamant that his team must grab the chance to shine at Japan/Korea 2002.

'To play in the World Cup is the highlight of a player's career,' he insists. 'And they should show just how much they want to be there.'

Already, Spanish clubs are impressing in Europe (Real Madrid – Champions League winners, 1998 and 2000; Valencia – Champions League runners-up, 2000 and 2001; Alaves – UEFA Cup runners-up 2001). Now it really is the time for the international side to do the same on the world stage.

MANAGER
JOSE ANTONIO CAMACHO

Since taking over from Javier Clemente in 1998, Camacho has breathed new life into Spain's internationals. The former Real Madrid defender has encouraged a more attacking and open style that appears to have gone down well with his players. It also makes for more attractive football than previously witnessed under Clemente.

ONE TO WATCH
RAUL

At just 17 years and four months, Raul Gonzalez Blanco became the youngest player ever to wear a Real Madrid shirt and quickly became a Spanish sporting idol. Although he didn't perform as well as hoped in France '98, the last few years have seen him emerge as one of the game's true talents and his performances will be crucial to Spain's progress in this year's finals.

SPAIN ESSENTIALS

◎ Spain's best showing in the World Cup came in Brazil in 1950 when they finished fourth.

THE ROAD TO WORLD CUP 2002

UEFA QUALIFYING GROUP 7

02/09/00	v Bosnia H (a)	won 2–1
07/10/00	v Israel (h)	won 2–0
11/10/00	v Austria (a)	drew 1–1
24/03/01	v Liech'stein (h)	won 5–0
02/06/01	v Bosnia H (h)	won 4–1
06/06/01	v Israel (a)	drew 1–1
01/09/01	v Austria (h)	won 4–0
05/09/01	v Liech'stein (a)	won 2–0

TOP SCORER
Raul, 4 goals

FINAL POSITION

P	W	D	L	F	A	GD	Pts	Psn
8	6	2	0	21	4	+17	20	1st

GROUP B FIXTURES

2 June	v Slovenia, Gwangju, 12.30pm
7 June	v Paraguay, Jeonju, 10am
12 June	v South Africa, Daejeon, 12.30pm

KOREA/JAPAN WORLD CUP 2002

GROUP A

FRANCE ☐	URUGUAY ☐
SENEGAL ☐	DENMARK ☐
31 May, Seoul, 12.30pm	1 June, Ulsan, 10am

FRANCE ☐	DENMARK ☐
URUGUAY ☐	SENEGAL ☐
6 June, Busan, 12.30pm	6 June, Daegu, 7.30am

DENMARK ☐	SENEGAL ☐
FRANCE ☐	URUGUAY ☐
11 June, Incheon, 7.30am	11 June, Suwon, 7.30am

GROUP B

PARAGUAY ☐	SPAIN ☐
SOUTH AFRICA ☐	SLOVENIA ☐
2 June, Busan, 8.30am	2 June, Gwangju, 12.30pm

SPAIN ☐	SOUTH AFRICA ☐
PARAGUAY ☐	SLOVENIA ☐
7 June, Jeonju, 10am	8 June, Daegu, 7.30am

SOUTH AFRICA ☐	SLOVENIA ☐
SPAIN ☐	PARAGUAY ☐
12 June, Daejeon, 12.30pm	12 June, Seogwipo, 12.30pm

GROUP C

BRAZIL ☐	CHINA ☐
TURKEY ☐	COSTA RICA ☐
3 June, Ulsan, 10am	4 June, Gwangju, 7.30am

BRAZIL ☐	COSTA RICA ☐
CHINA ☐	TURKEY ☐
8 June, Seogwipo, 12.30pm	9 June, Incheon, 10am

COSTA RICA ☐	TURKEY ☐
BRAZIL ☐	CHINA ☐
13 June, Suwon, 7.30am	13 June, Seoul, 7.30am

GROUP D

SOUTH KOREA ☐	UNITED STATES ☐
POLAND ☐	PORTUGAL ☐
4 June, Busan, 12.30pm	5 June, Suwon, 10am

SOUTH KOREA ☐	PORTUGAL ☐
UNITED STATES ☐	POLAND ☐
10 June, Daegu, 7.30am	10 June, Jeonju, 12.30pm

PORTUGAL ☐	POLAND ☐
SOUTH KOREA ☐	UNITED STATES ☐
14 June, Incheon, 12.30pm	14 June, Daejeon, 12.30pm

GROUP E

REP OF IRELAND ☐	GERMANY ☐
CAMEROON ☐	SAUDI ARABIA ☐
1 June, Niigata, 7.30am	1 June, Sapporo, 12.30pm

GERMANY ☐	CAMEROON ☐
REP OF IRELAND ☐	SAUDI ARABIA ☐
5 June, Ibaraki, 12.30pm	6 June, Saitama, 10am

CAMEROON ☐	SAUDI ARABIA ☐
GERMANY ☐	REP OF IRELAND ☐
11 June, Shizuoka, 12.30pm	11 June, Yokohama, 12.30pm

GROUP F

ENGLAND ☐	ARGENTINA ☐
SWEDEN ☐	NIGERIA ☐
2 June, Saitama, 10.30am	2 June, Ibaraki, 6.30am

SWEDEN ☐	ARGENTINA ☐
NIGERIA ☐	ENGLAND ☐
7 June, Kobe, 7.30am	7 June, Sapporo, 12.30pm

SWEDEN ☐	NIGERIA ☐
ARGENTINA ☐	ENGLAND ☐
12 June, Miyagi, 7.30am	12 June, Osaka, 7.30am

GROUP G

CROATIA ☐	ITALY ☐
MEXICO ☐	ECUADOR ☐
3 June, Niigata, 7.30am	3 June, Sapporo, 12.30pm

ITALY ☐	MEXICO ☐
CROATIA ☐	ECUADOR ☐
8 June, Ibaraki, 10am	9 June, Miyagi, 7.30am

MEXICO ☐	ECUADOR ☐
ITALY ☐	CROATIA ☐
13 June, Oita, 12.30pm	13 June, Yokohama, 12.30pm

GROUP H

JAPAN ☐	RUSSIA ☐
BELGIUM ☐	TUNISIA ☐
4 June, Saitama, 10am	5 June, Kobe, 7.30am

JAPAN ☐	TUNISIA ☐
RUSSIA ☐	BELGIUM ☐
9 June, Yokohama, 12.30pm	10 June, Oita, 10am

TUNISIA ☐	BELGIUM ☐
JAPAN ☐	RUSSIA ☐
14 June, Osaka, 7.30am	14 June, Shizuoka, 7.30am

All match times BST

WORLD CUP FIXTURES

Your compact guide to the biggest sporting event on the planet...

SECOND ROUND

MATCH A
Winner Group E
v Runner-up Group B
15 June, Seogwipo, 7.30am

MATCH B
Winner Group A
v Runner-up Group F
15 June, Niigata, 12.30pm

MATCH C
Winner Group F
v Runner-up Group A
16 June, Oita, 7.30am

MATCH D
Winner Group B
v Runner-up Group E
16 June, Suwon, 12.30pm

MATCH E
Winner Group G
v Runner-up Group D
17 June, Jeonju, 7.30am

MATCH F
Winner Group C
v Runner-up Group H
17 June, Kobe, 12.30pm

MATCH G
Winner Group H
v Runner-up Group C
18 June, Miyagi, 7.30am

MATCH H
Winner Group D
v Runner-up Group G
18 June, Daejeon, 12.30pm

QUARTER FINALS

QF1
Winner Match B
v Winner Match F
21 June, Shizuoka, 7.30am

QF2
Winner Match A
v Winner Match E
21 June, Ulsan, 12.30pm

QF3
Winner Match D
v Winner Match H
22 June, Gwangju, 7.30am

QF4
Winner Match C
v Winner Match G
22 June, Osaka, 12.30pm

SEMI-FINALS

SF1 Winner QF2 v Winner QF3
25 June, Seoul, 12.30pm

SF2 Winner QF1 v Winner QF4
26 June, Saitama, 12.30pm

THIRD / FOURTH PLACE PLAY-OFF

Loser SF1
v Loser SF2
29 June, Daegu, 12 noon

WORLD CUP FINAL

Winner SF1 v Winner SF2
30 June, Yokohama, 12 noon

SCORERS: SCORERS:

ENGLAND WORLD CUP 2002

DARK HORSES

CAMEROON

Their continent's most consistent team, Cameroon go to the World Cup Finals hoping to set a new benchmark for African soccer...

MANAGER
WINFRIED SCHÄFER
German-born coach Schäfer was formerly with Bundesliga clubs VfB Stuttgart and Karlsruhe. He ended months of bitter debate in Cameroon about the coaching position when he accepted the role. But he has his work cut out, having had only seven months prior to the tournament to get acquainted with his squad.

ONE TO WATCH
PATRICK MBOMA
Patrick Mboma emigrated to France with his family at the age of two and although he also owns a French passport, opted to play for his native Cameroon – although not until 1995, missing out on the 1994 World Cup. However, he did help his country reach France '98, scoring with five goals in the qualifiers, and in 2000 he was voted African Player of the Year.

CAMEROON ESSENTIALS

◉ Cameroon's best showing in the World Cup came at Italia '90 when they reached the quarter-finals, losing 2–3 to England.

THE ROAD TO WORLD CUP 2002

CAF AFRICAN QUALIFYING GROUP 1

18/06/00 v Libya (a)	won 3–0
09/07/00 v Angola (h)	won 3–0
28/01/01 v Togo (a)	won 2–0
25/02/01 v Zambia (h)	won 1–0
22/04/01 v Libya (h)	won 1–0
06/05/01 v Angola (a)	lost 0–2
01/07/01 v Togo (h)	won 2–0
14/07/01 v Zambia (a)	drew 2–2

TOP SCORER
Patrick Mboma, 6 goals

FINAL POSITION

P	W	D	L	F	A	GD	Pts	Psn
8	6	1	1	14	4	10	19	1st

GROUP E FIXTURES

1 June v Ireland, Niigata,	7.30am
6 June v S. Arabia, Saitama,	10.00am
11 June v Germany, Shizuoka,	12.30pm

Cameroon celebrate winning the 2000 Sydney Olympics

Cameroon's 'Indomitable Lions' are becoming something of a fixture at the World Cup Finals – this is their fourth consecutive appearance and their fifth in total, a record for the African continent. But they have yet to get past the quarter-finals and will be hoping they break through that barrier this time. First, though, they must qualify from a group that includes both Germany and the Republic of Ireland.

As England fans will recall, Cameroon first captured the imagination of the footballing world during Italia '90 with their charismatic blend of adventurous, attacking football and high spirits, personified by the the unique goal celebrations of 38-year-old Roger Milla. It was the first time an African team had played so confidently on the global stage and it was only in extra time in Naples that their run was finally ended by England.

This time, Cameroon fans are hoping to do even better than the last eight and set new landmarks for African football. And there are good reasons for optimism. They go into the tournament as reigning African champions and many of the current squad have already tasted international success with the Olympic triumph of Cameroon's under-23 side at the 2000 Games in Sydney. Key players from that surprise victory include the winger Lauren Etame Mayer, Samuel Eto'o and his strike partner, African Footballer of the Year Patrick Mboma.

They will be hoping to add their names to a growing list of Cameroon players who have made a name for themselves in the footballing world such as François Omam-Biyik, Thomas N'Kono and of course, Milla.

Cameroon could again be putting a smile on the faces of football fans the world over this summer.

CROATIA

They were the surprise package at the last World Cup and Croatia will be out to shock a few more people this time around...

The fitness of Alen Boksic could be key to Croatia's World Cup hopes

Few people predicted how well Croatia would perform at France '98. Third place is an achievement for most teams but especially so for a squad that were tournament fledglings.

A few new faces have been added to that squad but many familiar ones from 1998 remain, including Chelsea's Mario Stanic, Portsmouth's Robert Prosinecki, Stuttgart's Zvonimir Soldo, Juve's Igor Tudor, Inter's Dario Simic, Las Palmas' Robert Jarni and Igor Stimac of Hajduk Split.

Star striker Alen Boksic, who missed out on Euro '96 and France '98 with injuries, will also be an important player if he is fit to compete this time.

Among the newcomers are a number of players who also play in Europe's major leagues, such as Aston Villa striker Bosko Balaban, and Bayern Munich's brothers Robert and Niko Kovac.

Croatia slid down the world rankings after failing to qualify for Euro 2000, but new coach Mirko Jozic appears to have pointed the team in the right direction again. They breezed through qualifying for this World Cup and morale among the squad is high.

Jozic seems to have got the balance right between youth and experience and two consecutive World Cup Finals appearances isn't bad work for a side still in their infancy as an independent national squad. They could surprise a few people again this summer.

One player who is hoping to help them on their way is Davor Suker. The former Real Madrid star was club-less until a few months ago after quitting West Ham in the summer.

'My chances of playing in the World Cup are declining,' he admitted glumly. Since then, however, he has agreed terms with 1860 Munich and is confident he can do a job for Croatia.

'I am sure that I will prove my quality in the Bundesliga and be in good shape for the trip to Japan and South Korea,' said the 33-year-old, who will retire whatever happens this summer.

Croatia fans will be hoping he goes out with a bang.

MANAGER
MIRKO JOZIC

Following a legend isn't easy but that's what Jovic had to do when he took over the national side from Miroslav 'Ciro' Blazevic. However, the shadow of the previous incumbent has become less of a problem since a near flawless qualification that saw Jovic's team fly through eight matches undefeated.

ONE TO WATCH
ALEN BOKSIC

With Marseilles, Lazio and Juventus among the clubs on his CV, Boksic is one of Croatia's most celebrated players. At Lazio he won the Italian league title, although his best spell was at Marseilles where he became the country's top scorer, won the league championship and then went on to help Marseilles become the first French side ever to win the European Cup.

CROATIA ESSENTIALS

◎ Croatia's best showing in the World Cup came in France in 1998 when they finished third.

THE ROAD TO WORLD CUP 2002

UEFA QUALIFYING GROUP 6

02/09/00 v Belgium (a) drew 0–0
11/10/00 v Scotland (h) drew 1–1
24/03/01 v Latvia (h) won 4–1
02/06/01 v San Marino (h) won 4–0
06/06/01 v Latvia (a) won 1–0
01/09/01 v Scotland (a) drew 0–0
05/09/01 v San Marino (a) won 4–0
06/10/01 v Belgium (h) won 1–0

TOP SCORER
Bosko Balaban, 5 goals

FINAL POSITION

P	W	D	L	F	A	GD	Pts	Psn
8	5	3	0	15	2	+13	18	1st

GROUP G FIXTURES
3 June v Mexico, Niigata, 7.30am
8 June v Italy, Ibaraki, 10.00am
13 June v Ecuador, Yokohama, 12.30pm

German Number One Oliver Kahn barks out orders to his defenders

GERMANY

We beat them 5–1 in the qualifiers but Germany will be no pushovers when the finals tournament gets underway...

MANAGER
RUDI VÖLLER

The 1990 World Cup winning star who scored 47 goals in 90 internationals became Germany's caretaker coach after Germany's poor show under Erich Ribbeck at Euro 2000. When Bayer Leverkusen's Christoph Daum was ruled out following a drugs incident, Völler agreed to take the job on a permanent basis.

ONE TO WATCH
OLIVER KAHN

Since Andreas Köpke retired at the end of France '98, Oliver Kahn has established himself as Germany's Number One. Voted the European goalkeeper of the Year in 2000, the Bayern Munich star's performances will be a crucial factor for the Germans at Japan/Korea 2002.

WHEN GERMANY LOST TO England in Munich last September it was their worst home defeat for 70 years. Confidence drained, they followed up with a dismal goalless draw against Finland which meant they had to enter the last chance saloon of the play-offs. The way Rudi Völler's side were playing, a two-legged tie against the Ukraine appeared to be a tough obstacle, but displaying all the mental toughness of the great German sides, they prevailed courtesy of a 1–1 draw in Kiev and a 4–1 win in Dortmund.

Coach Rudi Völler deserves a lot of credit for their change in fortunes. He managed to lift his players' spirits very quickly after the painful defeat against Sven-Göran Eriksson's team and got his tactics absolutely spot on against the Ukraine. And the chants of 'Rudi, Rudi' which rang out inside the Westfalenstadion that night proved that the former German playing legend still has the backing of the fans.

Afterwards, their inspirational goalkeeper Oliver Kahn was in buoyant mood. 'I think we have great times ahead of us,' he said. 'I've always said that what we needed was a spark that would push us up to a higher level. Hopefully that was it.'

One only needs to glance at Germany's magnificent record in World Cup tournaments (which is second only to Brazil) to know that it is foolish to write them off. But it is true that since stars like Lothar Matthäus, Jürgen Klinsmann and Völler himself retired, the German national team has been in transition. If they are to emerge from this trough in June, the established Bayern Munich trio of Kahn, Mehmet Scholl and Jens Jeremies will require support from less experienced colleagues. Players such as Hertha Berlin creative midfielder Sebastian Deisler and free-scoring attacking midfielder Michael Ballack (who netted three in the play-offs against Ukraine) have to step up to the plate. And if Völler solves his striking problem – Oliver Bierhoff, Carsten Jancker, Oliver Neuville, Miroslav Klose and Gerald Asamoah have all been worryingly inconsistent – Germany might just prove the doubters wrong again.

GERMANY ESSENTIALS

Germany have won the World Cup three times (1954, 1974 and 1990) and reached the Final three times (1966, 1982 and 1986).

THE ROAD TO WORLD CUP 2002

UEFA QUALIFYING GROUP 9

02/09/00 v Greece (h) won 2–0
07/10/00 v Germany (a) won 1–0
24/03/01 v Albania (h) won 2–1
28/03/01 v Greece (a) won 4–2
02/06/01 v Finland (a) drew 2–2
06/06/01 v Albania (a) won 2–0
01/09/01 v England (h) lost 1–5
06/10/01 v Finland (h) drew 0–0

TOP SCORER
Michael Ballack, 6 goals

FINAL POSITION

P	W	D	L	F	A	GD	Pts	Psn
8	5	2	1	14	10	+4	17	2nd

Beat Ukraine 5–2 (agg) in play-offs to qualify

GROUP E FIXTURES
1 June v Saudi Arabia, Sapporo, 12.30pm
5 June v Rep Of Ireland, Ibaraki, 12.30pm
11 June v Cameroon, Shizuoka, 12.30pm

PORTUGAL

After their strong showing at Euro 2000, Portugal's star-studded squad are desperate to make an impact on the world stage...

Vision, acceleration and dribbling ability... Rui Costa has it all according to Eusebio

PORTUGAL ARE ENJOYING A golden era in their international football history. A third-place finish at Euro 2000 showed that they are far more than mere purveyors of beautiful football, and the team has gone from strength to strength since then.

Portugal went through a tricky-looking World Cup qualifying group unbeaten, racking up an extraordinary +26 goal difference in just 10 games. In Rui Costa and Luis Figo they boast two truly world-class players with the ability to unlock any defence and Nuno Gomes and Pauleta (seven goals each in the qualifiers) provide the cutting edge. No wonder then, that a London bookmaker immediately installed Portugal as 11/1 shots to win the 2002 World Cup when the draw placed them in a soft-looking group with co-hosts South Korea, Poland and the USA. But coach Antonio Oliveira is keen to ward off complacency.

'It's not as easy as it looks,' he claims. 'Korea are at home with big support and in Dutchman Guus Hiddink they have a top-class coach so they will be very well organised. Poland are a good European team and the USA have plenty of World Cup experience.'

Soccer legend Eusebio, who famously scored four goals against North Korea in the 1966 World Cup, has a more realistic view. 'We were lucky in the draw,' he admits. 'If we beat USA in our first game we will be halfway to the next phase. And after that it's in God's hands.'

Or perhaps at Figo's feet. The midfield superstar from Lisbon is the figurehead for a team that was born at youth level. Many of the side carrying Portuguese hopes at World Cup 2002 first played together in the team that won two consecutive FIFA World Youth Championships back in 1989 and 1991, finished runners-up in the 1994 UEFA Under-21s Championship and secured fourth position in the 1996 Olympic Games. There will be no more settled side at the World Cup and if Portugal play to their massive potential, they could go all the way.

MANAGER
ANTONIO OLIVEIRA

The first coach to take Portugal to a World Cup since 1986. He previously held the job between 1994 and 1996 and was recalled to replace Humberto Coelho who resigned following Portugal's semi-final defeat by France at Euro 2000. As manager of Porto, he won two league titles, in 1997 and 1998.

ONE TO WATCH
RUI COSTA

The AC Milan star is undoubtedly one of the finest midfielders in the world. As Portuguese legend Eusebio says, 'In Rui Costa, with his vision, acceleration and ability to dribble, Portugal have a creator of the highest order.' Costa and Figo will be Portugal's key playmakers.

PORTUGAL ESSENTIALS

◎ Portugal's best-ever performance in the World Cup was in 1966 when Eusebio inspired them to a third-place finish.

THE ROAD TO WORLD CUP 2002

UEFA QUALIFYING GROUP 2

03/09/00 v Estonia (a) won 3–1
07/10/00 v Rep of Ireland (h) drew 1-1
11/10/00 v Holland (a) won 2–0
14/02/01 v Andorra (h) won 3–0
28/03/01 v Holland (h) drew 2–2
02/06/01 v Rep of Ireland (a) drew 1-1
06/06/01 v Cyprus (h) won 6–0
01/09/01 v Andorra (a) won 7–1
05/09/01 v Cyprus (a) won 3–1
06/10/01 v Estonia (h) won 5–0

TOP SCORER
Nuno Gomes/Pauleta, 7 goals each

FINAL POSITION

P	W	D	L	F	A	GD	Pts	Psn
10	7	3	0	33	7	+26	24	1st

GROUP D FIXTURES
5 June v USA, Suwon, 10.00am
10 June v Poland, Jeonju, 12.30pm
14 June v S Korea, Incheon, 12.30pm

REST OF THE WORLD

BELGIUM

Scifo, Van Der Elst, Preud'homme... the World Cup just wouldn't be the same without the Belgians.
Indeed, not. They have now qualified for the final stages of six consecutive World Cup tournaments – a great achievement for a small nation. Mind you, they only just scraped in this time, beating the Czech Republic in the play-offs. A 1–0 home win and a late penalty by 'Mr 1000 Volts' in Prague sealed their place in the finals.

Erm, Mr 1000 Volts...
That's Marc Wilmots' nickname at his German club Schalke 04. Wilmots' running, tackling and long-range shooting have made him Belgium's most influential player since the great Enzo Scifo retired. The 33-year-old is the top scorer in the Belgian squad with 24 goals in 61 games – not bad for a midfielder.

Very impressive. Do their strikers score goals too?
Well, 23-year-old Emile Mpenza is the man most likely. He has averaged around a goal every three games since he broke into the international side back in 1997. Quick, powerful and two-footed, Mpenza's only problem now is shaking off a groin injury that has disturbed his progress in 2001–02.

How about the other players in the squad?
One thing they have is a breadth of experience, because since the Bosman ruling, the majority ply their trade abroad.

What are their chances of progressing in Japan?
Their group is relatively weak, so a second-round place is possible, but don't expect them to emulate the boys of '86 who reached the semi-finals in Mexico.

BELGIUM ESSENTIALS

THE ROAD TO WORLD CUP 2002

UEFA QUALIFYING GROUP 6
Date	Match	Result
02/09/00	v Croatia (h)	drew 0–0
07/10/00	v Latvia (a)	won 4–0
14/02/01	v San Marino (h)	won 10–1
24/03/01	v Scotland (a)	drew 2–2
02/06/01	v Latvia (h)	won 3–1
06/06/01	v San Marino (a)	won 4–1
05/09/01	Scotland (h)	won 2–0
06/10/01	v Croatia (a)	lost 0–1

TOP SCORER
Marc Wilmots, 7 goals

FINAL POSITION
P	W	D	L	F	A	GD	Pts	Psn
8	5	2	1	25	6	+19	17	2nd

Beat Czech Republic 2–0 (agg) in play-offs to qualify

GROUP H FIXTURES
4 June v Japan, Saitama, 10.00am
10 June v Tunisia, Oita, 10.00am
14 June v Russia, Shizuoka, 7.30am

Emile Mpenza: quick, powerful and two-footed... not sure about his vision though!

Veteran Fan Zhiyi will be a key player for China

CHINA

China. First time in the finals I believe...
Yes, after 44 years of waiting, the most populous nation on the planet have finally got a team to the World Cup. And they qualified in style, breezing through the Asian section under the guidance of experienced Serbian coach Bora Milutinovic.

A marvellous effort. But what sort of teams were they up against?
Good point. China's group was by far the easiest of the two Asian final qualifying groups. Beating teams like Uzbekistan, Qatar, Oman and the United Arab Emirates might not be the best preparation for taking on mighty Brazil in the first phase.

Who was that Chinese guy who played for Crystal Palace a year or two back?
You refer, I presume, to Fan Zhiyi who is now playing for Dundee in the Scottish Premier League. Zhiyi and his team-mate Sun Jihai joined Palace from Shanghai Shenua in 1998. Whereas Jihai has since returned home to China, Zhiyi established himself as one of the key players at Selhurst Park.

He's a defender, right?
Yes, that is 32-year-old Fan's preferred position, but his strength on the ball, speed and work-rate make him a powerful force in midfield too.

Any other players to look out for?
Like Zhiyi, veterans of previous near misses Hai Haidong and Ma Mingyu will be making the most of their first and last finals tournament. Of the younger generation Sun Jihai, Li Tie and Qu Bo all impressed during qualifying. Whatever happens now, China's mere qualification for the finals will give their FA confidence that their new professional league is starting to bear fruit.

CHINA ESSENTIALS

THE ROAD TO WORLD CUP 2002

AFC ASIAN QUALIFYING (SECOND PHASE) GROUP B
Date	Match	Result
25/08/01	v UAE (h)	won 3–0
31/08/01	v Oman (a)	won 2–0
07/09/01	v Qatar (a)	drew 1–1
15/09/01	v Uzbekistan (h)	won 2–0
27/09/01	v UAE (a)	won 1–0
07/10/01	v Oman (h)	won 1–0
13/10/01	v Qatar (h)	won 3–0
19/10/01	v Uzbekistan (a)	lost 0–1

TOP SCORER
Xie Hui, 7 goals

FINAL POSITION
P	W	D	L	F	A	GD	Pts	Psn
8	6	1	1	13	2	+11	19	1st

GROUP C FIXTURES
4 June v Costa Rica, Gwangju, 7.30am
8 June v Brazil, Seogwipo, 12.30pm
13 June v Turkey, Seoul, 7.30am

Hotshot Rolando Fonseca blasted 10 goals during the qualifiers.

COSTA RICA

Costa Rica? Don't recall seeing a lot of them in previous World Cups...
This is only the second time they have reached the finals (they first qualified in 1990) so they are pretty excited about it.

Oh. So what sort of style do they play?
They tend to go for a 3–5–2 formation, aiming to make their opponents go through the middle while they counter-attack on the flanks. Two wing-backs overlap constantly and two other midfielders stay back to fill in the gaps.

Right... does it work?
Well, they scored 25 goals in 14 qualifiers and since manager Alexander Guimares joined they've won 11, drawn six and only lost two matches.

Doesn't Paulo Wanchope play for Costa Rica?
He does, partnering Rolando Fonseca up front. They're the country's all-time leading scorers and mighty popular. Although not yet as popular as Hernan Medford, who played in the 1990 World Cup and scored the goal against Sweden that put Costa Rica in the second round. He's still in the team now and he pretty much *is* football in Costa Rican eyes.

Any other 'legends' I should know about?
Juan Ulloa... Javier Astua... ring any bells? If not, that's probably because the Costa Ricans aren't the best travellers. In fact they lost to Barbados away in qualifying. But, to be fair, they did also win at the Guillermo Caneda stadium in Mexico City – the first CONCACAF visiting team ever to do so.

So will Brazil be trembling in their boots?
Um, no, probably not.

COSTA RICA ESSENTIALS

THE ROAD TO WORLD CUP 2002

CONCACAF NORTH/CENTRAL AMERICA FINAL QUALIFYING ROUND

28/02/01 v Honduras (h) drew 2–2
28/3/01 v Trinidad & Tobago (h) won 3–0
25/04/01 v USA (a) lost 0–1
16/06/01 v Mexico (a) won 2–1
20/06/01 v Jamaica (h) won 2–1
01/07/01 v Honduras (a) won 3–2
01/09/01 v Trinidad & Tobago (a) won 2–0
05/09/01 v USA (h) won 2–0
07/10/01 v Mexico (h) drew 0–0
11/11/01 v Jamaica (a) won 1–0

TOP SCORER
Rolando Fonseca, 10 goals

FINAL POSITION

P	W	D	L	F	A	GD	Pts	Psn
10	7	2	1	17	7	+10	23	1st

GROUP C FIXTURES
4 June v China, Gwangju, 7.30am
9 June v Turkey, Incheon, 10.00am
13 June v Brazil, Suwon, 7.30am

DENMARK

So how are they doing since Schmeichel left?
Not bad. Sunderland's Thomas Sorensen had the unenviable task of filling the Great Dane's gloves but he's done pretty well. The Danes conceded less goals than any other side in their qualifying group.

Oh, I get it – they play a cagey defensive game, a bit like Norway?
Actually, they also scored more goals than their group rivals. In fact, they're very strong up front. They've got Schalke 04's Ebbe Sand, who was leading the scorers' table in the Bundesliga this season. And alongside him is Feyenoord's John Dahl Tomasson, who's a lot better than he looked at Newcastle.

Okay, so they've got a good 'keeper and some decent forwards. What about the midfield?
AC Milan's Thomas Helveg pulls the strings, with the help of Udinese's Martin Jorgensen. And Chelsea's Jesper Gronkjaer will be hoping to get a game too.

They sound quite useful...
They'll need to be. Senegal could cause an upset and who knows what Uruguay will come up with. Then there's the small matter of France...

Still, they should get through the group stage.
Yes, you'd fancy them to take second place in the group, but how far they will progress is anyone's guess. They've only made the last eight at the World Cup once. Mind you, that was last time around, at France '98, so if they're still on the up...

DENMARK ESSENTIALS

THE ROAD TO WORLD CUP 2002

UEFA QUALIFYING GROUP 3

02/09/00 v Iceland (a) won 2–1
07/10/00 v N. Ireland (a) drew 1–1
11/10/00 v v Bulgaria (h) drew 1–1
25/03/01 v Malta (a) won 5–0
28/03/01 v Czech Rep (a) drew 0–0
02/06/01 v Czech Rep (h) won 2–1
06/06/01 v Malta (h) won 2–1
01/09/01 v N. Ireland (h) drew 1–1
05/09/01 v Bulgaria (a) won 2–0
06/10/01 v Iceland (h) won 6–0

TOP SCORER
Ebbe Sand, 9 goals

FINAL POSITION

P	W	D	L	F	A	GD	Pts	Psn
10	6	4	0	22	6	+16	22	1st

GROUP A FIXTURES
1 June v Uruguay, Ulsan, 10.00am
6 June v Senegal, Daegu, 7.30am
11 June v France, Incheon, 7.30am

Bundesliga star Ebbe Sand is Denmark's main striking threat

REST OF THE WORLD

ECUADOR

Ecuador. Land of oil, emeralds and bananas...
Er, yes. And your point is?

Not really known for their football, are they?
No, but they finished second in the tough South American section (ahead of the likes of Brazil, Paraguay, Uruguay and Colombia) to qualify for the World Cup for the first time in the country's history. Things weren't looking too bright for them halfway through the campaign when they lost 0–4 away to Uruguay, but they fought back with six wins, three draws and just one defeat to claim an automatic place.

A decent side then...
Well, apart from deadly striker Agustin Delgado, Ecuador have a fairly modest set of players. But shrewd Colombian coach Hernan Dario Gomez has moulded a team that is more than the sum of its parts. Their excellent results in qualifying can't just be put down to the home altitude advantage of playing high up in the Andes.

Delgado? The guy who plays for Southampton
The very same. Delgado is the first player from Ecuador to grace the Premiership. The fact that he ended up level in the scoring charts on nine goals with Argentinian ace Hernan Crespo in the qualifiers shows his potential. 'El Tin' combines exceptional aerial ability with scintillating speed and he's a big-game player as indicated by his three goals in vital wins over Brazil (1–0) and Paraguay (2–1).

What does manager Gomez expect from his players in Japan?
Not too much. 'We are going to the World Cup to learn,' he says. Defender Ivan Hurtado agrees, but adds. 'Just because we are going to learn, doesn't mean we will comply. Just the opposite. We've earned the right to be there and we're going to compete.'

ECUADOR ESSENTIALS

THE ROAD TO WORLD CUP 2002

CONMEBOL SOUTH AMERICA QUALIFYING GROUP

29/03/00 v Venezuela (h)	won 2–0
26/04/00 v Brazil (a)	lost 2–3
03/06/00 v Paraguay (a)	lost 1–3
29/06/00 v Peru (h)	won 2–1
19/07/00 v Argentina (a)	lost 0–2
25/07/00 v Colombia (h)	drew 0–0
16/08/00 v Bolivia (h)	won 2–0
03/09/00 v Uruguay (a)	lost 0–4
08/10/00 v Chile (h)	won 1–0
15/11/00 v Venezuela (a)	won 2–1
28/03/01 v Brazil (h)	won 1–0
24/04/01 v Paraguay (h)	won 2–1
02/06/00 v Peru (a)	won 2–1
15/08/01 v Argentina (h)	lost 0–2
05/09/01 v Colombia (a)	drew 0–0
06/10/01 v Bolivia (a)	won 5–1
07/11/01 v Uruguay (h)	drew 1–1
14/11/01 v Chile (a)	drew 0–0

TOP SCORER
Agustin Delgado, 9 goals

FINAL POSITION

P	W	D	L	F	A	GD	Pts	Psn
18	9	4	5	23	20	+3	31	2nd

GROUP G FIXTURES
3 June v Italy, Sapporo, 12.30pm
9 June v Mexico, Miyagi, 7.30am
13 June v Croatia, Yokohama, 12.30pm

Saint Agustin Delgado

Ian Harte (left) and Robbie Keane celebrate the Republic's play-off win over Iran

IRELAND

So the Republic won a play-off at last...
Yes, after painful play-off losses in Euro '96, France '98 and Euro 2000, a 2–1 aggregate win over Iran got them back into the World Cup finals. Having negotiated the original 'Group Of Death' unbeaten, finishing on the same points as winners Portugal and knocking out Holland, it would've been a travesty if Mick McCarthy's boys hadn't made it.

Captain Roy Keane is a key figure for them, isn't he?
Manchester United fans have nicknamed him 'Massive' and with good reason. The most effective midfielder in the world, Keane is a fearsome tackler, metronomic passer and he also weighs in with important goals (four in the qualifiers). He leads by example and will not tolerate team-mates not putting in the same level of commitment. Just by his sheer presence he can lift the entire Irish side to heights previously undreamt of...

... Calm down, they aren't a one-man team.
Quite right. In goal, they've got Shay Given who saved them on numerous occasions against Iran and relishes the international stage. And the likes of Damien Duff, Matt Holland, Mark Kinsella and Mark Kennedy ably support Keane in midfield. Up front, the other Keane, Robbie, is emerging as a really potent striker.

Any chance of emulating the Italia '90 boys and making it to the last eight?
Definitely. The Republic have been drawn in one of the 'easier' groups and should they finish in the top two, they will probably play Spain or Paraguay in the second round.
'We know what we've got to do,' says boss McCarthy. 'Now the big challenge is for us to go and do it when it matters.'

REPUBLIC OF IRELAND ESSENTIALS

THE ROAD TO WORLD CUP 2002

UEFA QUALIFYING GROUP 2

02/09/00 v Holland (a)	drew 2–2
07/10/00 v Portugal (a)	drew 1–1
11/10/00 v Estonia (h)	won 2–0
24/03/01 v Cyprus (a)	won 4–0
28/03/01 v Andorra (a)	won 3–0
25/04/01 v Andorra (h)	won 3–1
02/06/01 v Portugal (h)	drew 1–1
06/06/01 v Estonia (a)	won 2–0
01/09/01 v Holland (h)	won 1–0
06/10/01 v Cyprus (h)	won 4–0

TOP SCORER
Roy Keane, 4 goals

FINAL POSITION

P	W	D	L	F	A	GD	Pts	Psn
10	7	3	0	23	5	+18	24	2nd

Beat Iran 2–1 (agg) in UEFA/AFC play-offs to qualify for the finals.

GROUP E FIXTURES
1 June v Cameroon, Niigata, 7.30am
5 June v Germany, Ibaraki, 12.30pm
11 June v S. Arabia, Yokohama, 12.30pm

Japanese football is getting stronger all the time

JAPAN

Let's face it, Japan are only in it because they're the hosts.
Well, that's a little harsh. After all, they almost qualified for USA '94 – if they hadn't conceded a last-minute goal against Iraq in their final qualifying match, they would have been there. And there have been other near misses.

But when they did qualify, didn't they lose all their games?
Yes, it's true that at France '98, they did suffer: their record was played three, lost three, but that was before they appointed Philippe Troussier as coach.

Who?
Frenchman Philippe Troussier. He coached Nigeria and Burkina Faso with some success and he's moulded the current team into a pretty competent unit. They won the Asian Cup in 2000 and the reached the final of the Confederations' Cup in 2001. Like Arsène Wenger, Troussier is a believer in youth and most of his side have been playing together since 1999, when they reached the final of the FIFA World Youth Championship in Nigeria.

So Japanese football has really improved?
Take our word for it, it has. Thanks mainly to the advent of the J League which has crucially made football popular with youngsters. More kids playing the game means more talent to choose from. And the quality of Japanese football is now high enough for players to be poached for other leagues. Junichi Inamoto earned a move to Arsenal and Hidetoshi Nakata plies his trade at Parma.

Nakata. I've heard of him. He must be pretty good.
You could say. He won Asian Footballer of the Year in 1998, the youngest player to ever win the award. His goals were the key reason Japan got to France '98. And despite the disappointing results for the team there, Nakata still did enough on the world stage to get European clubs interested. Perugia bought him for £2.3 million and he scored two goals on his debut. Roma paid £13 million for him a couple of years later and last year Parma forked out a massive £18.5 million for him. He could be worth even more by the end of this summer...

JAPAN ESSENTIALS

THE ROAD TO WORLD CUP 2002

Not applicable – Japan qualify automatically as joint hosts

GROUP H FIXTURES
4 June v Belgium, Saitama, 10.00am
9 June v Russia, Yokohama, 12.30am
14 June v Tunisia, Osaka, 7.30am

KOREA

Korea have been in the World Cup finals quite a few times, haven't they?
Five actually, 2002 will be their sixth appearance.

A good track record, then?
Um, not exactly. They've never actually won a game in the Finals. You might be thinking of North Korea who made the quarter-finals in 1966.

Never won a game? Can't be too happy about that.
They're not. Which is why last year they appointed the former PSV Eindhoven coach Guus Hiddink, the first non-Korean to take charge of the national side.

A touch of the Svens...
Indeed. Hiddink is under a bit of pressure to get Korea to the second round at least for the first time ever. But the draw hasn't been too unkind and he does have some useful players in his squad. including defender Hong Myung-bo, Yoo Sang-chul in midfield and, best of all, Seol Ki-Hyeon at his disposal.

Bit useful, is he?
Definitely. And a thinker too. Spotted in Korea's college league, Seol turned down a move into the J League because he wasn't sure he'd develop quickly enough there. Instead he decided to go to Europe. Antwerp took him on and he so impressed that when coach Regi van Accker moved to Lierse he invited Seol to go with him. But Seol again refused, taking the more difficult option of trying to impress at Anderlecht. Which he has done with some aplomb. His performances will probably dictate whether or not Korea reach stage two.

KOREA ESSENTIALS

THE ROAD TO WORLD CUP 2002

Not applicable – Korea Republic qualify automatically as joint tournament hosts

GROUP D FIXTURES
4 June v Poland, Busan, 12.30pm
10 June v USA, Daegu, 7.30am
14 June v Portugal, Incheon, 12.30pm

Seol Ki-Hyeon: a scholar and a scorer

REST OF THE WORLD

Blanco's boots

MEXICO

How many times have Mexico reached the finals?
Twelve. By tradition, Mexico have been the super power of the CONCACAF federation...

Conker what?
CONCACAF – it's the North and Central American World Cup qualifying region. Anyway, as I was saying, Mexico are always a sure-fire bet to make the finals, but there were some managerial casualties along the way this time.

Ooh, sounds messy.
It was. Manuel Lapuente resigned before the end of the semi-final round. Enrique Meza stepped in but was promptly sacked when he became the first Mexican manager to lose a qualifier on home soil. The 1986 World Cup player Javier Aguirre then took the reins and the Mexicans' form finally improved enough to qualify in second position behind Costa Rica.

What's their style of play?
A simple 4–4–2 formation. The defence is organised by Mexico's most capped player Claudio 'El Emparador' Suarez, and Aguirre has built a strong midfield around Alberto Garcia Aspe, a veteran of two World Cups. Up front, Cuauhtemoc Blanco is renowned for the party piece he revealed at France '98.

Eh?
You know, the 'Cuauhtémina' where he rests the ball on both feet and skips past opponents. It will be fun to see him again... if he turns up. The temperamental striker has threatened to quit the team in a row over airline tickets. Don't ask...

MEXICO ESSENTIALS

THE ROAD TO WORLD CUP 2002

CONCACAF NORTH/CENTRAL AMERICA FINAL QUALIFYING ROUND

28/02/01 v USA (a) lost 0–2
25/03/01 v Jamaica (h) won 4–0
25/04/01 v Trinidad & Tobago (a) drew 1–1
16/06/01 v Costa Rica (h) lost 1–2
20/06/01 v Honduras (a) lost 1-3
01/07/01 v USA (h) won 1–0
02/09/01 v Jamaica (a) won 2–1
05/09/01 v Trinidad & Tobago (h) won 3–0
07/10/01 v Costa Rica (a) drew 0–0
11/11/01 v Honduras (h) won 3–0

TOP SCORER
Cuauhtemoc Blanco, 9 goals

FINAL POSITION

P	W	D	L	F	A	GD	Pts	Psn
10	5	2	3	16	9	+7	17	2nd

GROUP G FIXTURES
3 June v Croatia, Niigata, 7.30am
9 June v Ecuador, Miyagi, 7.30am
13 June v Italy, Oita, 12.30pm

PARAGUAY

You say Paraguay, I say José Luis Chilavert!
Oh yes, one of the stars of France '98, the 36-year-old Paraguay goalkeeper makes Fabien Barthez look introverted. Not only is José a class 'keeper, he also takes Paraguay's penalties and the odd free-kick. He has scored more than 50 goals during his career (in qualifying, he hit the net four times!). We won't be seeing him for a while though...

Oh, why not?
José got sent-off against Brazil and he will be suspended for the first two games.

What did he do?
He spat at Roberto Carlos – hardly the behaviour of a wannabe politician. Then again...

Can Paraguay cope without him?
Well, Paraguay are renowned for their miserly defence and they have a very settled squad so maybe. Uruguayan coach Sergio Markarian led them to the second phase of the 1998 World Cup and he has steered them to the 2002 finals without making too many changes.

Okay, a miserly defence, but can they score goals?
They racked up an impressive 29-goal tally in qualification and the goals were shared around the team. Argentina coach Marcelo Bielsa rated them as 'the toughest rivals we faced in the qualifying', so Spain and co better not take them lightly.

PARAGUAY ESSENTIALS

THE ROAD TO WORLD CUP 2002

CONMEBOL SOUTH AMERICA FINAL QUALIFYING GROUP

29/03/00 v Peru (a) lost 0–2
26/04/00 v Uruguay (h) won 1–0
03/06/00 v Ecuador (h) won 3–1
29/06/00 v Chile (a) lost 1–3
18/07/00 v Brazil (h) won 2–1
27/07/00 v Bolivia (a) drew 0–0
16/08/00 v Argentina (a) drew 1–1
02/09/00 v Venezuela (h) won 3–0
07/10/00 v Colombia (a) won 2–0
15/11/00 v Peru (h) won 5–1
28/03/01 v Uruguay (a) won 1–0
24/04/01 v Ecuador (a) lost 1–2
02/06/00 v Chile (h) won 1–0
15/08/01 v Brazil (a) lost 0–2
05/09/01 v Bolivia (h) won 5–1
07/10/01 v Argentina (h) drew 2–2
08/11/01 v Venezuela (a) lost 1–3
14/11/01 v Colombia (h) lost 0–4

TOP SCORER
José Cardoso, 6 goals

FINAL POSITION

P	W	D	L	F	A	GD	Pts	Psn
18	9	3	6	29	23	+6	30	4th

GROUP B FIXTURES
2 June v S Africa, Busan, 8.30am
7 June v Spain, Jeonju, 10.00am
12 June v Slovenia, Seogwipo, 12.30pm

The goalscoring goalkeeper: José-Luis Chilavert unleashes another trademark free-kick

POLAND

When did Poland last qualify for a World Cup?
In 1986. And apart from a silver medal at the 1992 Olympics, the Poles haven't had a great deal to celebrate for a while.

They used to be pretty good though, didn't they?
Yes, they beat Brazil to take third place in the 1974 Finals and beat France to take third in 1982 as well. But it's been a while since they qualified for either the World Cup or European Championships.

Now they're back. Why?
It's really down to one player – Nigerian-born Emmanuel Olisadebe. He took Polish citizenship and has been scoring for them ever since. He hit the target seven times as Poland lost only once in the qualifying stages. They topped their group comfortably and now Olisadebe has a chance to shine on the world stage. The fans love him.

I bet they do. But are they just a one-man team?
No, the defender Tomasz Hajto is also vital to the side. He's not only an excellent defender but he also has one of the longest throws in football – even longer than Gary Neville's. It wreaks havoc in the opposition's penalty area. The only problem with Hajto, who plays his club football in the Bundesliga, is that he has a tendency to get booked. In the 1998–99 season alone, he picked up 16 yellow cards. That was more than anyone else in the division. To be fair, though, he's not a dirty player, just a hard tackler. Nonetheless, Poland will be hoping he doesn't pick up any suspensions if they are to progress in the tournament.

POLAND ESSENTIALS

THE ROAD TO WORLD CUP 2002

UEFA QUALIFYING GROUP 5

Date	Match	Result
02/09/00	v Ukraine (a)	won 3–1
07/10/00	v Belarus (h)	won 3–1
11/10/00	v Wales (h)	drew 0–0
24/03/01	v Norway (a)	won 3–2
28/03/01	v Armenia (h)	won 4–0
02/06/01	v Wales (a)	won 2–1
06/06/01	v Armenia (a)	drew 1–1
01/09/01	v Norway (h)	won 3–0
05/09/01	v Belarus (a)	lost 1–4
06/10/01	v Ukraine (h)	drew 1–1

TOP SCORER
Emmanuel Olisadebe, 7 goals

FINAL POSITION

P	W	D	L	F	A	GD	Pts	Psn
10	6	3	1	21	11	+10	21	1st

GROUP D FIXTURES
4 June v Korea, Busan, 12.30pm
10 June v Portugal, Jeonju, 12.30pm
14 June v USA, Daejeon, 12.30pm

Nigerian-born striker Emmanuel Olisadebe has transformed Polish fortunes

Spot the Spartak Moscow players...

RUSSIA

Ah, the old Soviet Union.
That's the problem. They're not the Soviet Union any more and they can't draw players from a number of Soviet republics.

Is that why they weren't in the last World Cup?
It's difficult to say. The last Soviet team to compete was in the 1990 Finals, but Russia did qualify for USA '94. And although they didn't make it to France in 1998, they did top their qualifying group this time.

Anyone in particular to look out for?
Alexandre Mostovoi and Dmitri Khoklov get a few goals for the team but the name currently in the spotlight is their lethal striker Vladimir Beschastnykh.

Beschastnykh? Must be a big spotlight.
Ho ho. Beschastnykh scored a hat-trick in the game against Switzerland which saw Russia qualify. Hence all the fuss about him. But others have performed well too, such as goalkeeper Rouslan Nigmatoulline, who let in just five goals in 10 qualifiers.

Who's the coach?
Oleg Romantsev. He's also the coach of Spartak Moscow and appears to have taken a leaf out of the old Soviet bosses' book by building his international team around his club side. Beschastnykh is just one of several Spartak players in the team. Romantsev can also call on Celta Vigo's Mostovoi and Valery Karpin. And he has the experienced Viktor Onopko, who played in the '94 Finals.

RUSSIA ESSENTIALS

THE ROAD TO WORLD CUP 2002

UEFA QUALIFYING GROUP 1

Date	Match	Result
02/09/00	v Switzerland (a)	won 1–0
11/10/00	v Luxembourg (h)	won 3–0
24/03/01	v Slovenia (h)	drew 1–1
28/03/01	v Faroe Islands (h)	won 1–0
25/04/01	v Yugoslavia (a)	won 1–0
02/06/01	v Yugoslavia (h)	drew 1–1
06/06/01	v Luxembourg (a)	won 2–1
01/09/01	v Slovenia (a)	lost 1–2
05/09/01	v Faroe Islands (a)	won 3–0
06/10/01	v Switzerland (h)	won 4–0

TOP SCORER
Vladimir Beschastnykh, 7 goals

FINAL POSITION

P	W	D	L	F	A	GD	Pts	Psn
10	7	2	1	18	5	+13	23	1st

GROUP H FIXTURES
5 June v Tunisia, Kobe, 7.30am
9 June v Japan, Yokohama, 12.30pm
14 June v Belgium, Shizuoka, 7.30am

REST OF THE WORLD

SAUDI ARABIA

The Irish should thrash this lot, shouldn't they?
You obviously don't know your Asian football, mate. Saudi Arabia have been the Persian Gulf's major force for the past two decades. Since 1984 they have won a record-equalling three Asian Cup titles and this will be their third consecutive appearance at the World Cup finals. At USA '94, the 'Sons of the Desert' became the first Asian team to reach the second round since North Korea in '66.

Did they qualify easily?
Actually it was all rather dramatic. They needed to win their last game against Thailand then hope Bahrain beat group leaders Iran. Amazingly, Bahrain upset the odds to beat Iran 3–1 and the Saudis' 4–1 win over Thailand clinched the one automatic qualifying spot.

Who's the Saudi manager?
God knows! Since 1996, there have been no fewer than eight changes of Head Coach, with Otto Pfister and Nasser Al Johar both holding the position twice. At the time of writing, the position is still up for grabs.

Hmm, I think I'll give it a miss. Who are the players to watch?
Talal Al Meshal, Obeid Al Dossary and Sami Al Jaber (who was once on loan to Wolves) were their main goalscorers in qualifying with 11, nine and eight goals respectively. And there are also high hopes for Asian Player of The Year Nawaf Al Temyat who missed qualifying with a serious knee injury but should be back for the finals.

SAUDI ARABIA ESSENTIALS

THE ROAD TO WORLD CUP 2002

AFC ASIAN QUALIFYING (SECOND PHASE) GROUP A

17/08/01 v Bahrain (h) drew 1–1
24/08/01 v Iran (a) lost 0–2
31/08/01 v Iraq (h) won 1–0
15/09/01 v Thailand (a) won 3–1
21/09/01 v Bahrain (a) won 4–0
28/09/01 v Iran (h) drew 2–2
05/10/01 v Iraq (a) won 2–1
21/10/01 v Thailand (h) won 4–1

TOP SCORER
Talal Al Meshal, 11 goals

FINAL POSITION

P	W	D	L	F	A	GD	Pts	Psn
8	5	2	1	17	8	+9	17	1st

GROUP E FIXTURES
1 June v Germany, Sapporo, 12.30pm
6 June v Cameroon, Saitama, 10.00am
11 June v Rep Of Ire, Yokohama, 12.30pm

The 'Sons Of The Desert' will be no walkover for the Irish

El Hadji Diouf (right) scored two hat-tricks in the qualifiers

SENEGAL

Senegal, the surprise package of the qualifiers...
Yes, these World Cup first-timers (the only one of the five African representatives who have never been to the finals before) were not expected to make it. But they finished top of arguably the toughest group in the continental qualifiers ahead of Morocco, Egypt and Algeria.

The only Senegalese star I know is Patrick Vieira... who doesn't play for them.
No, but Senegal has a strong French connection. Coach Bruno Metsu is French and the vast majority of the squad are based in France while the others play their club football in European leagues such as Austria, Greece, Switzerland and Portugal.

Who's their best player?
Powerful 21-year-old Rennes striker El Hadji Diouf (pictured) was a revelation in the qualifiers scoring nine of Senegal's 14 goals, including two hat-tricks. Adoring Senegal fans have bestowed upon him the rather dubious nickname 'Serial Killer' for his prolific scoring ability.

Who are Senegal's other key players?
Diouf has formed a useful partnership with Auxerre striker Khalilou Fadida, who almost elected to play for Belgium before winning his first cap for Senegal in 2000. And width is provided by Henri Camara while defender Ferdinand Coly organises things at the back.

How was their World Cup qualification greeted back in Senegal?
There were wild celebrations. President Adboulaye Wade even cut short a state visit to fly home and join in the fun with the players and public. Bruno Metsu promises that his side will bring charm to the finals. 'Senegal will be the Jamaica of this World Cup,' he says.

SENEGAL ESSENTIALS

THE ROAD TO WORLD CUP 2002

CAF AFRICAN QUALIFYING GROUP 3

16/06/00 v Algeria (a) drew 1–1
09/07/00 v Egypt (h) drew 0–0
24/02/01 v Morocco (a) drew 0–0
10/03/01 v Namibia (h) won 4–0
21/04/01 v Algeria (h) won 3–0
06/05/01 v Egypt (a) lost 0–1
14/07/01 v Morocco (h) won 1–0
21/07/01 v Namibia (a) won 5–0

TOP SCORER
El Hadji Diouf, 9 goals

FINAL POSITION

P	W	D	L	F	A	GD	Pts	Psn
8	4	3	1	14	2	+12	15	1st

GROUP A FIXTURES
31 May v France, Seoul, 12.30pm
6 June v Denmark, Daegu, 12.30pm
11 June v Uruguay, Suwon, 7.30am

Coach Srecko Katanec and star player Zlatko Zahovic celebrate qualifying for World Cup 2002

SLOVENIA

Slovenia haven't been around long, have they?
No, quite. Formerly a part of Yugoslavia (until 1991) they only have around 2 million inhabitants but have managed to forge together a football team of remarkable resilience that continues to improve.

They're on the up, then?
You could say. By the end of the qualifying campaign they had risen to 26th in the FIFA world rankings, a leap of some 100 places from when the system began in 1993. Considering that Slovenia mustered just one point in the qualifying stages of the 1998 World Cup, they've come a long way.

Who did they beat to qualify this time?
They were drawn in UEFA Group 1 with Russia, Yugoslavia, Switzerland and the Faroe Islands and, after finishing undefeated but in second place with five wins and five draws, they beat Romania 3–2 in the play-offs.

What's their secret?
It's pretty simple really. Coach Srecko Katanec believes in hard work, team spirit and solid defending. In Slovenia he's considered something of a hero.

What about the players – anyone stand out?
Zlatko Zahovic is the playmaker. A cult figure in Slovenia, even though he's never played his club football there, he's an energetic attacking midfielder who would fit in well in most of Europe's international teams. A great passer and excellent tackler, he also averages a goal every two games.

Sounds almost too good to be true...
Well, he's also injury prone and a little temperamental which has caused spats with coach Katanec. But he is the key to Slovenia's success this summer.

SLOVENIA ESSENTIALS

THE ROAD TO WORLD CUP 2002

UEFA QUALIFYING GROUP 1

03/09/00 v Faroe Islands (a)	drew 2–2
07/10/00 v Luxembourg (a)	won 2–1
11/10/00 v Switzerland (h)	drew 2–2
24/03/01 v Russia (a)	drew 1–1
28/03/01 v Yugoslavia (h)	drew 1–1
02/06/01 v Luxembourg (h)	won 2–0
06/06/01 v Switzerland (a)	won 1–0
01/09/01 v Russia (h)	won 2–1
05/09/01 v Yugoslavia (a)	drew 1–1
06/10/01 v Faroe Islands (h)	won 3–0

TOP SCORER
Zlatko Zahovic, 4 goals

FINAL POSITION

P	W	D	L	F	A	GD	Pts	Psn
10	5	5	0	17	9	+8	20	2nd

Beat Romania 3-2 (agg) in play-offs to qualify

GROUP B FIXTURES
2 June v Spain, Gwangju, 12.30pm
8 June v South Africa, Daegu, 7.30am
12 June v Paraguay, Seogwipo, 12.30pm

SOUTH AFRICA

South Africa. They're another team that haven't been on the international circuit that long...
No, that's true. But their long period of isolation – due to apartheid – has been over for a decade now and in the last ten years they've grown in strength as a footballing force.

How good are they?
They've been the top African nation in the FIFA rankings for some time and qualified for France '98. They also won the African Nations' Cup in 1996 and with more and more of their players competing in European leagues, they're likely to continue improving.

Are they good enough to get through their group?
Definitely. They had two draws and a defeat in France '98 and they'll be expecting to do better than that this time. Players Sibusiso Zuma, Delron Buckley and Siyabonga Nomvete are all potential match winners.

Anyone I've heard of?
What about Shaun Bartlett? The Charlton striker is fast, good in the air and has excellent ball control. The South Africans think so highly of him they've made him captain (Nelson Mandela even went to his wedding!).

Who's in charge?
Coach Carlos Queiroz, who lead Portugal to the World Under-20 Championship in 1991. He did a good job in the qualifiers too as South Africa easily topped their group. Corporate sponsorship in sport and South Africa's economic strength means there are funds available to encourage further development in football. But a success this summer would be even more beneficial.

SOUTH AFRICA ESSENTIALS

THE ROAD TO WORLD CUP 2002

CAF AFRICAN QUALIFYING GROUP 5

09/07/00 v Zimbabwe (a)	won 2–0
27/01/01 v Burkino Faso (h)	won 1–0
25/02/01 v Malawi (a)	won 2–1
05/05/01 v Zimbabwe (h)	won 2–1
01/07/01 v Burkino Faso (a)	drew 1–1
14/07/01 v Malawi (h)	won 2–0

TOP SCORER
Shaun Bartlett 4 goals

FINAL POSITION

P	W	D	L	F	A	GD	Pts	Psn
6	5	1	0	10	3	+7	16	1st

GROUP B FIXTURES
2 June v Paraguay, Busan, 8.30am
8 June v Slovenia, Daegu, 7.30am
12 June v Spain, Daejeon, 12.30pm

Charlton star striker and South African captain Shaun Bartlett leads his team out

REST OF THE WORLD

TUNISIA

Tunisia, our first victims at the 1998 World Cup finals...
Yes, an Alan Shearer header and a corker by Paul Scholes earned us a comfortable 2–0 win.

Has their team changed much since?
Not really. The core of the current side was originally put together at under-23 level to compete at the 1996 Olympics, went on to play at France '98 and four years on has developed into a very efficient unit. Veterans like goalkeeper-captain Chokri El Quaer and Adel Sellimi are complemented by the youthful talent of Hassen Gabsi, Ali Zitouni and Tunisia's top scorer in qualifying Ziad Jaziri.

I hear they finished top of their qualifying group...
You can't read too much into that. Their opposition – Ivory Coast, Congo and Congo DR, Madagascar – hardly constitute the cream of world football. It wasn't a completely smooth ride though because their coach Francesco Scoglio left during the campaign. German Eckhard Krautzen replaced Scoglio and guided Tunisia safely to the finals, but promptly resigned at the conclusion of the qualifiers.

What's their World Cup history?
Tunisia's main claim to fame is that they were the first African side ever to win a match at football's biggest tournament, beating Mexico 3–1 at the 1978 World Cup finals in Argentina. This will be only their third finals appearance, and although Tunisia boasts the strongest league clubs in Africa, their achievements in international football are minimal compared to the likes of Cameroon and Nigeria.

TUNISIA ESSENTIALS

THE ROAD TO WORLD CUP 2002

CAF AFRICAN QUALIFYING GROUP 4

18/06/00 v Ivory Coast (a)	drew 2–2
08/07/00 v Madagascar (h)	won 1–0
28/01/01 v Congo (a)	won 2–1
25/02/01 v Congo DR (h)	won 6–0
05/05/01 v Madagascar (a)	won 2–0
20/05/01 v Ivory Coast (h)	drew 1–1
01/07/01 v Congo (h)	won 6–0
15/07/01 v Congo DR (a)	won 3–0

TOP SCORER
Ziad Jaziri, 6 goals

FINAL POSITION

P	W	D	L	F	A	GD	Pts	Psn
8	6	2	0	23	4	+19	20	1st

GROUP H FIXTURES
5 June v Russia, Kobe, 7.30am
10 June v Belgium, Oita, 10.00am
14 June v Japan, Osaka, 7.30am

Tunisia veteran Adel Sellimi battles for the ball

Turkey annihilated Austria 6–0 in the play-offs to claim their place at World Cup 2002

TURKEY

Just scraped through to the finals via the play-offs I believe...
Hmm, if you can describe a 6–0 aggregate win over Austria as 'scraping through'. Turkey actually had a very strong qualifying campaign, losing just one out of 10 group matches, and it was only undefeated Sweden's exceptional form that denied them top spot.

But what's Turkey's form like at major tournaments?
This will be their first appearance at the World Cup finals since 1954, but they have reached the quarter-finals of the last two European Championships so it's about time they made their mark on the global stage.

Turkey's coach must be a popular guy...
Oh no, the Turkish media have consistently attacked Senol Gunes as a dour figure with neither the experience nor charisma to lead Turkey. Many have called for the colourful Fatih Terim to return to the post of national coach. Unmoved, Gunes has urged all Turks to get behind him and his team. 'We don't need heroes and villains,' he proclaims. 'We are building a revolution here. We are soldiers and we are ready. Turkey is the best!'

Very rousing. What would he regard as success?
Gunes has stated that his priority is to qualify from the group and reach the quarter-finals which doesn't seem unreasonable. He has a strong squad, largely drawn from Istanbul clubs Galatasaray, Fenerbahçe and Besiktas. The rest such as defender Alpay Ozalan (who scored a hat-trick in qualifying against Macedonia) and midfielder Muzzy Izzet play at the top level of other European leagues. And if star striker Hakan Sukur hits form, they could do some real damage.

TURKEY ESSENTIALS

THE ROAD TO WORLD CUP 2002

UEFA QUALIFYING GROUP 4

02/09/00 v Moldova (h)	won 2–0
07/10/00 v Sweden (a)	drew 1–1
11/10/00 v Azerbaijan (a)	won 1–0
24/03/01 v Slovakia (h)	drew 1–1
28/03/01 v Macedonia (a)	won 2–1
02/06/01 v Azerbaijan (h)	won 3–0
06/06/01 v Macedonia (h)	drew 3–3
01/09/01 v Slovakia (a)	won 1–0
05/09/01 v Sweden (h)	lost 1–2
07/10/01 v Moldova (a)	won 3–0

TOP SCORER
Hakan Sukur, 6 goals

FINAL POSITION

P	W	D	L	F	A	GD	Pts	Psn
10	6	3	1	18	8	+10	21	2nd

Beat Austria 6–0 (agg) in play-offs to qualify

GROUP C FIXTURES
3 June v Brazil, Ulsan, 10.00am
9 June v Costa Rica, Incheon, 10.00am
13 June v China, Seoul, 7.30am

Claudio Reyna... the best American player ever?

The Americans are never going to master 'soccer' are they?
Hang on a second. They did reach the World Cup semi-finals in 1930 and, er, they won the CONCACAF Gold Cup in 1991. They've also been known to beat England on occasion...

Hmph. The USA women's team is much better...
They have been successful, yes, but the knock-on effect is that the game is more popular than it's been for some time in the States. And since France '98, when they finished last in their group, there's been a much-needed injection of youth which has bolstered the team no end.

Who's idea was that?
Manager Bruce Arena, who has won practically every honour going in the States. He started bringing in players from all over the world, giving 20 new players caps. Traditionally, the team has always been all right defensively but now they've got some class in midfield and up front too.

Like who?
Winger Ernie Stewart plies his trade for NAC Breda in Holland. He scored eight goals during the qualifiers. And the skipper Claudio Reyna is at Rangers.

Oh, Reyna is quite good.
He's arguably the best player the United States have ever had. His dad played professionally in Argentina. Reyna would have been at the Finals in 1994 but for injury. Nonetheless, he was still snapped up by Bayer Leverkusen and later moved on to Wolfsburg where he became the first American to captain a Bundesliga team. A great playmaker, he moved on to Rangers in 1999. Incidentally, his wife Danielle played for the US Women's team...

Perhaps they should get his missus in this summer...
Stop that.

USA ESSENTIALS

THE ROAD TO WORLD CUP 2002

CONCACAF NORTH/CENTRAL AMERICA FINAL QUALIFYING ROUND

28/02/01 v Mexico (h)	won 2–0
28/03/01 v Honduras (a)	lost 1–2
25/04/01 v Costa Rica (h)	won 1–0
16/06/01 v Jamaica (a)	drew 0–0
20/06/01 v Trinidad & Tobago (h)	won 2–0
01/07/01 v Mexico (a)	lost 0–1
01/09/01 v Honduras (h)	lost 2–3
05/09/01 v Costa Rica (a)	lost 0–2
07/10/01 v Jamaica (h)	won 2–1
11/11/01 v Trinidad & Tobago (a)	drew 0–0

TOP SCORER
Ernie Stewart, 8 goals

FINAL POSITION

P	W	D	L	F	A	GD	Pts	Psn
10	5	2	3	11	8	+3	17	3rd

GROUP D FIXTURES

5 June v Portugal, Suwon, 10.00am	
10 June v Korea, Daegu, 7.30am	
14 June v Poland, Daejeon, 12.30pm	

Didn't Uruguay used to be quite good once?
You could say that – they won the World Cup twice.

That is good...
Yeah, although the triumphs of 1930 and 1950 are some way behind them now. They'd only reached two of the last six Finals before this summer.

Are they back, back, back?
Steady on. Some have described them as a sleeping giant about to reawaken but that's probably a little fanciful. However, since taking over from Daniel Passarella, coach Victor Pua has maintained a steady course towards this year's finals.

So they qualified easily?
Well, after some inconsistent results, they clinched fifth place in the South American qualifying group (beating Brazil and drawing with Argentina along the way) to earn a play-off against Australia, which they won 3–1 on aggregate.

So they can turn it on when they want to...
They certainly have the players. Paolo Montero at Juventus, Alvaro Recoba at Internazionale, Pablo Garcia at Milan and Gonzalo De los Santos at Valencia, to name but a few.

Who's the key player?
Recoba, without doubt. The fans claim he has magic in his left boot and on his shoulders rests the responsibility of ensuring Uruguay get past the first round. The last time they did, in 1970, they finished fourth.

URUGUAY ESSENTIALS

THE ROAD TO WORLD CUP 2002

CONMEBOL SOUTH AMERICA FINAL QUALIFYING GROUP

29/03/00 v Bolivia (h)	won 1–0
26/04/00 v Paraguay (a)	lost 0–1
03/06/00 v Chile (h)	won 2–1
28/06/00 v Brazil (a)	drew 1–1
18/07/00 v Venezuela (h)	won 3–1
26/07/00 v Peru (h)	drew 0–0
15/08/00 v Colombia (a)	lost 0–1
03/09/00 v Ecuador (h)	won 4–0
08/10/00 v Argentina (a)	lost 1–2
15/11/00 v Bolivia (a)	drew 0–0
28/03/01 v Paraguay (h)	lost 0–1
24/04/01 v Chile (a)	won 1–0
01/07/01 v Brazil (h)	won 1–0
14/08/01 v Venezuela (a)	lost 0–2
04/09/01 v Peru (a)	won 2–0
07/10/01 v Colombia (h)	drew 1–1
07/11/01 v Ecuador (a)	drew 1–1
14/11/01 v Argentina (h)	drew 1–1

TOP SCORER
Dario Silva, 5 goals

FINAL POSITION

P	W	D	L	F	A	GD	Pts	Psn
18	7	6	5	19	13	+6	27	5th

Uruguay beat Australia 3–1 (agg) in the Oceania/CONMEBOL play-off to qualify

GROUP A FIXTURES

1 June v Denmark, Ulsan, 10.00am	
6 June v France, Busan, 12.30pm	
11 June v Senegal, Suwon, 7.30am	

We're going to the World Cup! Alvaro Recoba celebrates Uruguay's play-off win over Australia

THE MANAGER

THE GAFFER

He's led England from the brink of World Cup disaster to qualification for the finals and now SVEN-GÖRAN ERIKSSON is all set for the big one...

SVEN-GÖRAN ERIKSSON IS RACING against time. Shortly, he's due to catch a plane to Manchester to watch Manchester United v Liverpool, but he's kindly managed to squeeze me in for a chat about England's World Cup prospects before he sets off. However, when I'm shown into his office at The F.A.'s Soho Square headquarters, he appears to be just as calm as his public persona would suggest. He greets me with a smile and settles back into his chair before answering my questions. Our conversation is occasionally broken by the buzz of an internal telephone call. 'Don't worry, we'll make it...' I hear him reply before unhurriedly reclaiming his train of thought. Clearly, whether he's on the bench or talking to the journalists, Sven doesn't do 'flustered'.

That composure must have been useful when he attended the World Cup draw in Korea last December. A gasp of disbelief echoed around the Bexco Convention Centre in Busan when Argentina, Nigeria, Sweden and England were drawn together in Group F, but how did Sven react?

'At the time, I just hoped we would have more luck in the summer, because that night it couldn't have been worse,' he laughs. 'We drew the most difficult teams from every group. Argentina are the best team in the world. Sweden are one of the strongest in Europe, and Nigeria are considered the best team in Africa.'

'But when I thought about taking this job I dreamed about playing against teams like Argentina in the World Cup Finals. Maybe I had not envisaged playing them in the first phase, but while it is a difficult task, it is certainly not an impossible one.'

And the England coach is keen to emphasise that our prospective group opponents will be just as wary of us.

'I'm sure the managers of Sweden, Argentina and Nigeria felt the same way as me when they heard the draw. Before I went to Busan, I had heard that a lot of teams didn't want to meet England. Once we beat Germany 5–1 away, I think people changed their attitude towards us. In my opinion, we have top class players. Players such as Beckham and Owen could play for any of the top clubs or national sides in the world.'

Talking of Argentina and David Beckham, I ask Sven how he thinks the England captain will cope with what is sure to be billed as a revenge clash by the tabloid press.

'I have no worries at all,' he asserts. 'I didn't know David in 1998, but since I've been working with him I've been impressed by his maturity as a player and as a person. If he can score a goal like the one he did against Greece to get his team to the World Cup finals, he can deal with anything. I think mentally, he showed something very special.'

It's obvious that David Beckham has thrived since becoming England captain and while many people doubted Becks' leadership potential, he has proved an inspired choice. So how does their relationship work?

'Sometimes I phone him, or he phones me,' he says. 'I don't speak to him about everything, but if I have something I want to say to the team, I phone David and he will then talk to the rest of the players."

Judging by England's spectacular revival under Eriksson, his approach to management really works. As England Under–21 coach David Platt (who played under Eriksson at Sampdoria) maintains: 'Sven's good at man-management, group management and tactics. He is quiet, but not aloof and can be subtly inspirational. He is a definite winner.'

But can the nation's favourite Swede achieve the ultimate success of leading England to victory at World Cup 2002?

'Our first target is to reach the second phase,' he replies. 'It will be very difficult to win the tournament but we should try. Why can't we at least dream about it?'

The telephone buzzes again. This time he really must go.

'I've never been to a World Cup before as the coach of a national team and I've always wanted to be there,' he says as he escorts me to reception. 'Ever since I was a 10-year-old watching the 1958 World Cup Finals on TV it's been my great ambition to be involved. I'm looking forward to it very much.'

> 'It will be very difficult to win the tournament but... why shouldn't we dream about it?'

SVEN-GÖRAN
ERIKSSON

THE CHEF

EAT TO WIN!

Top footballers have to watch their diets, but luckily for the England lads top chef ROGER NARBETT will be on hand in Japan to make sure the menus never get boring…

IN 1989, REPRESENTATIVES of The F.A. International Committee travelled out to Albania to check the facilities before England's World Cup qualifier in the Eastern bloc country. They identified a serious problem – the food supply. Fearing a potential link between stomach and football upsets, it was decided that the England team would need to take their own chef. Roger Narbett got the job.

'At the time, I was in partnership with my father at the Bell Inn pub–restaurant near Lilleshall,' explains the 41-year-old chef. 'The Committee often met there and they enjoyed the food, so my name cropped up. I just was in the right place at the right time.'

Chef Roger Narbett (far right, top row) celebrates with the lads.

Roger is being rather too modest, because having learnt his culinary trade under the famous Roux brothers at The Waterside and Le Gavroche and held executive sous chef positions at swanky hotels such as The Dorchester, he was more than qualified.

Today, in his thirteenth year as the England squad's official chef, I'm meeting Roger at his own gastro-pub, the Bell & Cross in Holy Cross, Worcestershire.

'At the start, I only went on the trips to Eastern European countries,' he says as he places a delicious-looking chicken fillet in front of me. 'But it's progressed to every away trip and tournament. It's important to maintain consistency for the players.'

Roger is making meticulous preparations to achieve this aim at World Cup 2002. Once the team's training base in Japan is confirmed, he will travel out to meet the hotel's chef and the Conference & Banqueting manager. And come May, Roger himself will be in the kitchen cooking and guiding the hotel's kitchen staff.

So what culinary delights can the players look forward to in Japan?

'In a warmer climate, players tend to go more for pasta, soups, salads and lighter desserts like fruit and yoghurt,' he enthuses. 'The players like their rice and the fish over there is fantastic too. I only take a few essentials and I'll use fresh, local produce.

'On average, you are looking to give them about 60 per cent carbohydrates, 25 per cent fat and 15 per cent protein in their meals. You push the carbos more as the game approaches with pasta, fish, white meats etc. Then straight after the match, you begin to refuel them with more protein. If you've some time before the next match, red meat and chips are okay. There's a time for everything. But three days before a game you have to limit fat intake. At a tournament, it's very important to get the right balance.'

Roger should know, he's designed the menus for England at the European Championship in 1992 and 2000, Le Tournoi '97 and France '98, and Becks and co now enjoy a greater variety of food than ever before.

'In the early days, I'd take all the food with me and the players got a set menu,' he recalls. 'I would take things especially for certain players – for instance, Peter Shilton couldn't do without his pre-match rice pudding and Steve McMahon loved his Porridge Oats – but now I do a buffet style menu so there's always something for everyone.'

He laughs when I ask what sort of feedback he gets from the players:

'They give me a bit of stick if they're not happy! But it's all good banter. I've grown up with a lot of them and whether you're the coach driver or the chef, the players make you feel part of the team.

'I'm really looking forward to Japan – it'll be a great experience to be there together, trying to win the World Cup for everyone back home.'

And judging by my grilled lemon chicken, England certainly won't fail for the want of delicious, healthy food.

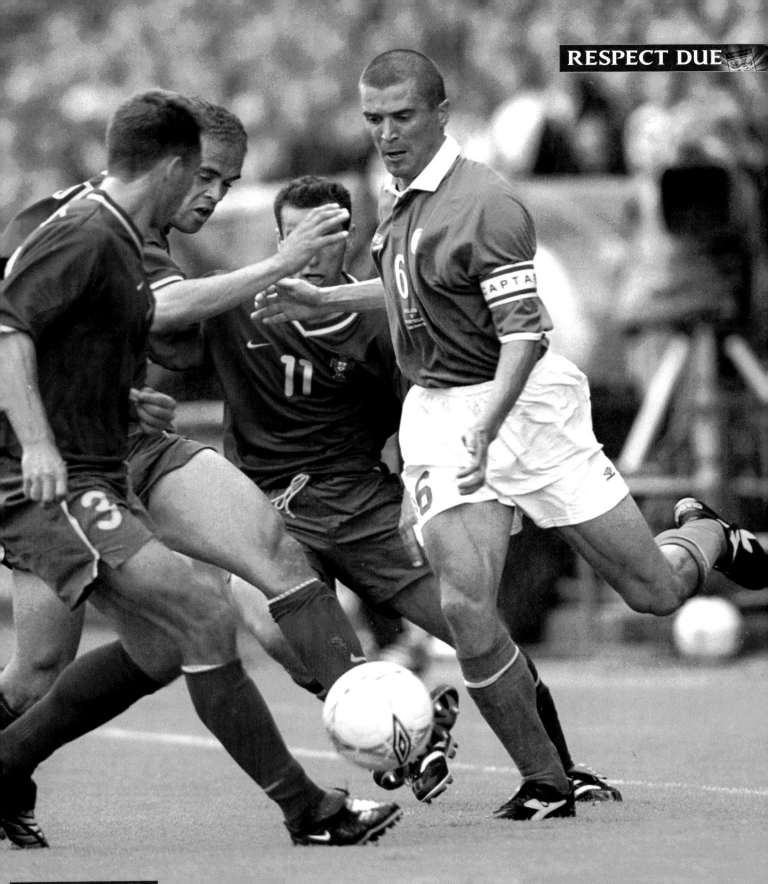

RESPECT DUE

ROY KEANE — REP. OF IRELAND

Ireland's World Cup hopes may hinge on the performances of their inspirational captain. As his manager Mick McCarthy says: 'For sheer effectiveness in a football match, Roy can't be bettered. He'd be my first pick ahead of Rivaldo, Figo, Zidane... any of them. This man never has a bad game – he has good games and ones that are even better.'

LIONHEARTS

MIDFIELD MAGICIAN

'PAUL SCHOLES, he scores goals...' goes the terrace chant and the midfielder has certainly poached a few crucial ones for England...

'I've always tried to play the same way for England as I do for my club'

PAUL SCHOLES HATES INTERVIEWS. Paul Scholes likes to keep himself to himself. Paul Scholes lets his football do the talking. Paul Scholes shuns the limelight so attractive to some of his peers...

As you may have gathered by now, there's a bit of a formula to writing a feature on Paul Scholes. Although to be fair to the nation's writers, they have a point. The Manchester United and England star has never been particularly fond of talking about himself ('My worst nightmare' is how he once described the process). Today however, he has at least combined our chat with something he infinitely prefers – spending time at home with his family. Hence we're sitting in the kitchen discussing football and – because of regular interjections from Scholes jnr – fatherhood.

Despite the scenes of domestic bliss, it soon becomes clear that Scholes has lost none of his hunger for success on the pitch.

'I'm my biggest critic,' he admits. 'If I know I've not played well, football doesn't get spoken about when I get home. I know when I've done well and when I haven't and a bad game stays on my mind.'

Not that he has a lot of bad games. A players' player, Scholes doesn't need to grab the headlines to gain the respect of the footballing world. His Manchester United team-mates even voted him their 2000–01 Player Of The Year. Naturally, Paul didn't get carried away.

'Of course it's nice to get praise from your team-mates,' he says bashfully. 'Nice that they think you did well, but football's a team game. That's what I'm more concerned with.'

It's a shame Paul doesn't enjoy praise because there's been plenty of it flying around during his career. From Johan Cruyff describing him as a 'a truly gifted player' to Patrick Vieira describing him as 'the best English player', the 5' 7" attacking midfielder has attracted admiration all along the way.

'I try not to pay too much attention and I do get a little bit embarrassed,' admits Paul. Best not mention what his former England colleague Tony Adams said about him then:

'Because Paul hasn't got the high profile of some players he doesn't get too much attention but he is one very fine footballer. He's an intelligent player, he works hard and he scores some great goals. He is not flamboyant and is a quiet lad off the pitch but he is a tremendous asset to England.'

A fact that has been proved time and time again. When England have needed a breakthrough goal, more often than not it's been Scholes who has provided it. Yet despite such frequent match-saving endeavours, he continues to avoid the limelight.

'That's just the way I like it,' he smiles. 'I come into training, work hard then go home. Everybody's different aren't they? But that's the way I like it and so far it's worked for me.'

Both for club and country. Paul is one of the few Premiership players who seem capable of regularly replicating their club form for their country. According to Scholes, the reason for his international success is simple:

'I've always tried to play the same way for my country as I do for my club. I think the problem some people may have had is when they try and change their normal game for internationals. When I was first picked for England, I was told to "play the way you do at United" and that's what people should do, instead of trying to be something different.'

Our club and country discussion is interrupted by an appearance from Paul's son Aaron, who needs to share some important information with his dad. 'Hmm,' nods Paul attentively, 'you'd like to do the interview for me?'

I think he's joking, but hastily move on to my last question. As Scholes jnr kicks a ball around the kitchen, I ask Paul if he feels his own footballing talents were in the genes.

'I'm not sure. My dad played quite a lot but never to a high standard,' he says of his father, Stewart. 'I don't know where my football ability has come from. I'm not sure it's come from anyone. Maybe it's just something that comes naturally. Obviously I did have a lot of help from my parents, and my dad came to all my games, but I don't think they drove me on. He'd say the odd thing now and again but that was it. I've just always liked football and always just wanted to play the game.'

Fortunately for all England fans, he does it exceedingly well.

ENDURING IMAGE

CAPTAIN MARVEL

16 June 1982, Bilbao
World Cup Finals Group Match
England 3 France 1

England skipper Bryan Robson celebrates scoring one of the fastest goals in World Cup Finals history. Robbo struck with a header just 27 seconds into England's opening match of Spain '82.

THE KNOWLEDGE

Thrill / bore* your friends at the pub with your amazing / scary* knowledge of essential / pointless* England World Cup trivia! (*delete as appropriate)

1. 1966 World Cup hero Bobby Moore became the only player ever to appear as captain both for and against England when he skippered Team America against England in 1976.

2. In 1986, Gary Lineker fired in six goals and became the only British player ever to finish as top scorer in a World Cup Finals tournament.

3. England played seven consecutive games without defeat in the 1966 and 1970 World Cup finals, including six successive wins, just one short of Italy's record set in 1934–38. The run ended when Brazil beat England 1–0 in the group stage of the Mexico World Cup.

4. Cameroon striker Roger Milla became the oldest player ever to play against England in a World Cup Finals tournament. Milla was 38 years and 37 days young when he came on as sub in the 1990 quarter-final match.

5. Alongside the great Brazilian goalkeeper Gylmar (who first achieved the feat in 1958), Gordon Banks and Peter Shilton share the all-time record for consecutive clean sheets in World Cup finals matches. Both Banksie (in 1966) and Shilts (1982) kept the opposition scoreless for four consecutive matches.

6. Three English referees – George Reader (1950), Bill Ling (1954) and Jack Taylor (1974) – have taken charge of a World Cup Final.

7. Only five England players have scored in two separate World Cup tournaments. They are: Tom Finney (1954 & 1958), Bobby Charlton (1962 & 1966), Geoff Hurst (1966 & 1970), Martin Peters (1966 & 1970), and Gary Lineker (1986 & 1990).

8. During Italia '90, England were captained by three different players. Bryan Robson initially wore the armband but sustained an achilles tendon injury and was sent home after the second group game. After that Peter Shilton and defender Terry Butcher both had turns as skipper on the way to England's elimination in the semi-final.

9. Manchester City goalkeeper Joe Corrigan is the heaviest player ever to be picked for an England World Cup squad. Big Joe was a trim 14st 12lbs for Spain '82, having weighed in at a blubbermongous 15 stone 11lbs for his England debut in 1976.

10. Billy Wright is the only English player to captain England in three separate World Cup finals. Wright led the team at the 1950, 1954 and 1958 tournaments.

11. Geoff Hurst (1966 v West Germany) and Gary Lineker (1986 v Poland) are the only England players ever to score hat-tricks in a World Cup tournament.

12. Midfielder Neil Webb came off the bench to play his last-ever game for England in the 1990 World Cup Third Place play-off match against Italy, but his first international sub appearance was far more noteworthy. On 9 September 1987 Webb came on to play against Germany in Dussledorf to become the 1000th footballer to be capped by England.

13. No one has managed England in more World Cup finals tournaments than Walter Winterbottom. Winterbottom was the manager in 1950, 1954, 1958 and 1962.

14. England's Peter Shilton kept more clean sheets (10) in World Cup finals matches than any other goalkeeper in history. That's one safe pair of hands!

15. Seven English footballers have played in three World Cup finals competitions. They are: Billy Wright (1950/54/58), Tom Finney (1950/54/58), Bobby Charlton (1962/66/70), Bobby Moore (1962/66/70), Terry Butcher (1982/86/90), Bryan Robson (1982/86/90) and Peter Shilton (1982/86/90).

16. The XI chosen by Bobby Robson to play against the Republic Of Ireland at Italia '90 was the most experienced England starting line-up ever. Shilton, Stevens, Pearce, Robson, Walker, Butcher, Waddle, Gascoigne, Lineker, Barnes and Beardsley had a combined total of 563 caps between them.

17. Peter Shilton played more World Cup finals matches (17) than any other England footballer. Bobby Charlton, Bobby Moore and Terry Butcher are equal second on 14 while Gary Lineker is third on 12 matches.

18. The 1966 World Cup Final is the only England match ever to feature goals scored in both the last minute of normal time (Weber for Germany) and extra time (Hurst for England). The last England player to score in the final minute of a World Cup finals match was David Platt in the 120th minute against Belgium in 1990.

19. Midfielders Ray Wilkins (6 June 1986 v Morocco) and David Beckham (30 June 1998 v Argentina) are the only two England players ever to be sent off in World Cup finals matches.

20. Bobby and Jack Charlton are the only brothers ever to be both selected for an England World Cup squad (1966 & 1970). Gary and Phil Neville almost emulated the Charltons in 1998, but although Phil played in the warm-up games, he was omitted from the final 22.

21. 1966 World Cup-winning manager Alf Ramsey also represented England as a player appearing at right back in all three of England's matches in the 1950 tournament.

22. England's all-time record in World Cup finals matches is:

P	W	D	L	F	A	GD	%wins
45	20	12	13	62	42	+20	44.4%

RESPECT DUE

FRANCESCO TOTTI — ITALY

Hailed by Sir Alex Ferguson as the best player in the world, the Roman maestro is like an Italian version of Peter Beardsley. Totti's ability on the ball and visionary passing could easily propel a typically solid-looking Italian team to glory.

TAKING ON THE WORLD

THE QUEST FOR GLORY

'History Will Teach Us Nothing,' sang Sting once. We don't agree. Let's look back at half a century of England seeking football's ultimate prize...

ALTHOUGH THE INAUGURAL WORLD Cup was staged in 1930, England didn't enter the competition until the first tournament after World War II, the 1950 event hosted in Brazil. By which stage there had already been three winners: Uruguay (1930), Italy (1934) and Italy again (1938). Securing a place in the 1950 finals for England was based not on the now more traditional qualifying group but instead on the Home International Championship matches with Scotland, Wales and Northern Ireland.

1950 BRAZIL

WITH MUCH OF EUROPE STILL suffering from the aftermath of the war, the choice of South America as hosts seemed a sensible decision. England, after qualifying wins of 4–1, 9–2 (!) and 1–0 against Wales, Northern Ireland and Scotland respectively, were named joint favourites for the tournament along with the hosts Brazil. English football was buoyant. Attendances were at an all-time high, and the national team had put together a string of impressive wins, including two victories over reigning world champions Italy and a 10–0 thrashing of Portugal in Lisbon.

The 1950 tournament was on a league basis and England, as one of four seeded teams, were drawn with Chile, the United States and Spain. The first two of which were unknown quantities.

Nonetheless, England made a reasonable start with a 2–0 win over Chile. The scoreline was perhaps a little flattering to a patchy English performance that drew boos from some of the South American fans. However, it was as good as things got for England.

Four days later we faced the United States. Against a backdrop of international drama – America was on the verge of getting embroiled in the Korean War – England suffered their most infamous international defeat to date. Going down 1–0 to a goal by Haitian centre-forward Joe Gaetjens sent shockwaves through English football and made headlines throughout the footballing world. Ironically though, in the States, where football was not considered a major sport, reports were tucked away in the small print.

England's next game, against Spain, proved to be our last of the tournament as we lost 1–0 to Spain. It was a disappointing end to the country's first World Cup venture. The team flew home and Uruguay went on to win the Jules Rimet trophy.

USA striker Joe Gaetjens shocks the football world by scoring the winner against England

Nat Lofthouse out-jumps Uruguay's Maspoli, but England went down to a 4–2 quarter-final defeat by the two-time World champions

1954 SWITZERLAND

AT THE HEIGHT OF THE COLD WAR, the choice of the politically neutral Switzerland as hosts was another prudent one. England again qualified through Home internationals. Although this time the players didn't travel with the wildly jingoistic 'kings of football' tag. The disappointments of 1950 together with a crushing pre-tournament 7–1 defeat by Hungary ensured a high level of pessimism accompanied them into the tournament. The organisers had also abandoned the league system for an altogether more layered and complicated arrangement which meant seeded teams in the same group didn't play each other. Draws were not desired and extra time was to be played if teams were level after 90 minutes. The ensuing confusion aided no one.

England's first match was a case in point. Facing Belgium, the teams finished level at 3–3 after normal time (England letting a 3–1 advantage slip). A further 30 minutes saw both teams score again to leave all players exhausted and a draw still the final result.

A 2–0 victory over hosts Switzerland in our next match was enough to ensure England topped the table (we never had to face the other seeded team in their group, Italy, who instead ended up playing Switzerland twice and still went out).

A draw and a win meant England progressed to the quarter-final where we faced two-times tournament winners Uruguay. Goals from our star players Nat Lofthouse and Tom Finney were not enough to stop the South Americans running out 4–2 winners. England were out. Uruguay went the same way in their semi-final against Hungary, who were themselves beaten in the final by West Germany.

'The disappointments of 1950 plus a crushing 7–1 defeat by Hungary ensured a high level of pessimism about England's chances at Switzerland '54.'

1958 SWEDEN

TO QUALIFY FOR THE FINALS IN Sweden, England topped a three-team group featuring Denmark and the Republic of Ireland. But the Munich Air Disaster, which tragically robbed Manchester United of some of their finest players, also had repercussions for the England team – Roger Byrne, Tommy Taylor, David Pegg and Duncan Edwards were the names no longer available to grace the tournament.

So it was a re-jigged England team which arrived in Sweden to face Austria, the Soviet Union and favourites Brazil. They came back from 0–2 down to draw 2–2 with the Soviet Union in the first game and performed well above expectations in the next match against Brazil. However, despite a total of 27 shots (14 from Brazil, 13 from England), the match ended 0–0. Another draw, 2–2 against Austria, meant England had to face the Soviet Union again, in a play-off to determine who qualified from the group alongside Brazil. A goal from the Soviet striker Ilyin ensured it wasn't England. Brazil went on to win the tournament.

TAKING ON THE WORLD

1962 CHILE

Bobby Moore and England win the World Cup after a memorable match against West Germany at Wembley

THREE WINS AND A draw again saw England qualify for the World Cup, just as they had in '58. This time Portugal and Luxembourg were the teams overcome and Chile was the destination reached.

Although the South Americans were hardly in a state of rude health to act as hosts – severe earthquake damage had affected around a third of all buildings in the country – they managed to stage the tournament.

Hungary, Argentina, and Bulgaria were the first-round opponents England faced, and a 2–1 defeat to Hungary was hardly the ideal start. The English press lambasted the team, claiming they were technically inept, tactically naive, and almost certainly bound for a first-round exit from the tournament.

The response from the players was to run Argentina ragged, running out 3–1 winners and securing our first win at the finals since 1954. And a 0–0 draw against Bulgaria five days later meant a place in the quarter-finals. As in 1954, England's place in the last eight meant a tie against the reigning champions. This time, the mighty Brazil. With the scores level at 1–1 at half-time, England looked like a team for whom the last four was a real possibility but second-half goals from Garrincha and Vava killed off the dream. Brazil went on to retain the trophy.

'As in 1954, England's place in the last eight meant a tie against the reigning champions – this time we faced the mighty Brazil.'

1966 ENGLAND

BACK IN 1960 FIFA HAD CONFIRMED that when the World Cup was next held in Europe, England would be the hosts. Our time had arrived and the nation knew there would never be a better chance to win football's greatest honour. But things didn't start too well as England huffed and puffed their way to a goalless draw against smooth-passing Uruguay. The match ended goalless but the bookies immediately upped England's odds of winning the trophy.

However, two 2–0 victories, over Mexico and France, ensured England topped the

'England were in the World Cup Final. And following the semi-final win over Portugal, there was a new wave of belief that we could play exciting quality football.'

group and kept the advantage of playing their games at Wembley. We had reached the quarter-finals for the third time and for the third time we faced South American opposition. This time it was Argentina.

England would also have to manage without Jimmy Greaves who had a gashed shin. The nation was distraught. In his place came Geoff Hurst, a 24-year-old West Ham striker with just five caps. However, 77 minutes into the quarter-final Hurst scored his first World Cup goal, a glancing header securing a 1-0 victory.

Manager Alf Ramsey stuck with the same team for the semi-final against Portugal and the players justified his faith with a breathtaking display. England ran out 2–1 winners thanks to a brace from Bobby Charlton, inspired defending from Nobby Stiles and superb saves from Gordon Banks.

England were in the World Cup Final and there was a new belief that we could play exciting, quality football. For four days a nation celebrated and anticipated.

Controversially, Ramsey stuck with the same team for a third consecutive game for the final against former winners West Germany, meaning no place for Greaves. Come matchday, there was further cause for nervousness as the Germans took an early lead after just 13 minutes. But both matters were addressed six minutes later as Hurst equalised. Peters made it 2–1 in the second half and England just had to hang on for 13 minutes. They managed 12, Weber equalising in the dying seconds to take the game to extra-time. England had been ten seconds away from the prize. Now they had to do it all again.

Ten minutes into extra time, Hurst did, hitting a shot against the underside of the bar. Did it bounce over the line? The officials said it did and England were in the lead. Hurst then ended any speculation by completing his hat-trick.

It was a moment to savour. England had won the World Cup. We were, for the next four years at least, the kings of football.

TAKING ON THE WORLD

1970 MEXICO

OIL HAD BEEN DISCOVERED IN THE North Sea, Concorde had made its maiden flight and man had walked on the moon but far more importantly, could England retain the World Cup? We got off to a reasonable start in Mexico, with a win against Romania but then had to face Brazil. Despite a spirited display (and some supernatural goalkeeping by Gordon Banks) England were defeated 1–0 in a tight match. However, a scrappy 1–0 victory over Czechoslovakia secured a place in the quarter-final against 1966's vanquished finalists, West Germany. The Germans were out for revenge and England, minus Gordon Banks (who had come down with a case of food poisoning), were taken to extra time after being held 2–2 after 90 minutes. Unlike '66, this time there would be no happy ending – it was Germany's Müller who got on the scoresheet in the additional period and England were out. Brazil went on to win the Jules Rimet trophy for the third time and as a reward were allowed to keep it.

> '2–2 after 90 minutes, but unlike '66 there was to be no happy ending'

1974 W. GERMANY

In 1974, there was a new World Cup trophy, but surprisingly England would not be among the nations competing for it in West Germany.

As a seeded team, we had the luxury of being drawn in a three rather than four-team qualifying group and our unseeded opponents Wales and Poland were not expected to pose any serious threat.

Alf Ramsey's team began with an unconvincing 1–0 win in Cardiff, but a 1–1 draw in the return at Wembley set off warning signals. It was the first time England had ever dropped a point at home in a World Cup qualifier. England then subsided to a 2–0 defeat in Poland (with Alan Ball sent off) meaning we simply had to win the return match at Wembley to qualify.

But an amazing performance by Polish goalkeeper Jan Tomaszewski frustrated England, who could only manage a 1–1 draw. We had failed to qualify and the glorious Ramsey World Cup era was over.

The greatest save ever? Gordon Banks dives low to his right to turn away Pele's header during England's group match against Brazil at Mexico '70

1978 ARGENTINA

Finland, Italy and Luxembourg were England's qualifying opponents but it was Italy who claimed the one available finals' slot. Under Don Revie and then Ron Greenwood, England beat both Finland and Luxembourg twice, but a 2–0 defeat by Italy in Rome was costly and goal difference cruelly denied the Three Lions.

1982 SPAIN

It was now twenty years since England had actually qualified for a World Cup (in 1966 we were hosts and in 1970 current holders of the trophy). This time, it looked achievable, with a qualifying group of Hungary, Romania, Switzerland and Norway. Nonetheless, it went to the wire and England only just scraped into second place with a 1–0 victory over group winners Hungary in our final match at Wembley. Paul Mariner was the scorer. Relief as much as celebration was the main reaction following the result.

England were drawn to play France, Czechoslovakia and Kuwait in the first stage of the finals and made a perfect start when Bryan Robson scored against France after just 27 seconds – one of the fastest goals in World Cup history. Robbo added a second after the interval and England ran out 3–1 winners.

Two more victories followed and England topped their group easily to qualify for the second stage.

The draw was not kind though and we now faced West Germany and hosts Spain. Two goalless draws were not enough to take us into the semi-finals and the players flew home with the small consolation of knowing they'd been undefeated in the tournament.

Italy won the final 3–1 against the Germans but England fans couldn't help but wonder what might have been if the draw had been a little kinder.

'Maradona's first goal may have earned him the label "cheat", but his second strike proved he was also a footballing genius.'

1986 MEXICO

An assured qualification instilled the England team with a certain amount of confidence going into the finals in Mexico but their first game didn't go to plan — a 1–0 defeat to Portugal. A scoreless draw with Morocco followed and victory over Poland became a necessity. Happily, Gary Lineker's hat-trick in a 3-0 win ensured it was achieved. Two more strikes from Lineker, this time against Paraguay, put England in the last eight. But England would progress no further in the competition because of one man: Diego Maradona. His 'hand of God' goal may have earned him the label 'cheat' but his second strike proved he was also a footballing genius, running the England defence ragged before stroking the ball past Shilton. Lineker pulled one back (making him the tournament's leading scorer) but England were out. Argentina, and Maradona, went on to take the trophy.

Lineker celebrates equalising against West Germany in the semi-final

1990 ITALY

Despite a fairly torrid time during qualification, England made it to Italia '90 and it was there that the real drama unfolded. Bobby Robson's team were drawn against the Republic of Ireland, Holland and Egypt. Two draws and a win ensured they topped the group. Next up were Belgium and David Platt's astonishing 360-degree turn and volley clinched a place in the last eight. There we faced Cameroon who'd beaten Argentina on the opening day of the finals with nine men. England, 2–1 down at one stage, dug deep and two clinical Gary Lineker penalties saw England through to the semi-finals for the first time on foreign soil. Standing between us and the final were old foes, West Germany.

The Germans took the lead after 60 minutes from a cruelly deflected free-kick but that man Lineker struck to level. Gascoigne was coming in for special attention – he had been fouled seven times – but when he lunged at Berthold he was booked for the second time in the tournament ensuring he would miss the final. His tears touched the nation as did the way he pulled himself together to continue England's struggle.

Extra time couldn't separate the teams and it went to penalties. Misses from Pearce and Waddle meant it would be the Germans in the final. They won it, but England's players returned home with heads held high.

'Gazza's tears touched the nation as did the way he pulled himself together to continue England's struggle.

That blond man (Ronald Koeman) should not have been on the pitch after this foul on Platt

1994 USA

Disappointment followed the heroics of Italia '90 as England failed to qualify for USA '94.

Graham Taylor had taken over as manager from Bobby Robson and England started well, consolidating a 1–1 draw against Norway with three straight wins against Turkey (twice) and San Marino, and scoring 12 goals without reply. But things started to go wrong when England surrendered a 2–0 lead to draw 2–2 with Holland at Wembley. Further 0–2 defeats in Norway and, controversially, in Holland (when Ronald Koeman should have been sent off for a cynical foul on David Platt but scored a brilliant free-kick two minutes later) followed.

So despite hammering seven past San Marino in our final qualifying match, England missed out on the finals for the third time in six attempts, as Norway and Holland took the top two places in UEFA Group 2.

Argentina 'keeper Roa saves David Batty's penalty and England are out of France '98

1998 FRANCE

In hindsight, the 1998 World Cup may be looked back on by England fans as being the making of two men. For very different reasons, the worlds of Michael Owen and David Beckham would never be the same after France '98.

As usual, England went into the tournament with no little expectation. Glenn Hoddle's side certainly had some potential stars, although Paul Gascoigne was not among their number, the manager having controversially left him out of the final squad. Neither Owen nor Beckham started England's first match, a 2–0 victory over Tunisia, although Owen did come off the bench. He did the same against Romania, this time scoring, but it wasn't enough to prevent a 2–1 defeat. Victory against Colombia became a neccessity, and a reality thanks to a superb Beckham free-kick and a great strike by Darren Anderton. England were in the second round but would have to overcome Argentina to progress. In a dramatic encounter, Owen scored the goal of the tournament, David Beckham became (temporarily) Public Enemy No.1 and England went out on penalties. France went on to win the trophy but Owen and Beckham would return, older, wiser and with world-class potential transformed into ability.

ROUGH GUIDE

FROM SAITAMA

England will play our matches in Japan, and the further we go in the tournament, the more travelling fans

GROUP PHASE MATCHES

SAITAMA
Our first stop on the road to World Cup glory is the Saitama Prefecture in the Greater Tokyo metropolis for the Sweden game on 2 June.

Situated just to the north of downtown Tokyo, Saitama lives up to its billing as 'The Colourful Prefecture'. For sheer natural beauty, the breathtaking Nakatsu-kyo Gorge with its 100-metre cliffs can't be beaten and green-fingered fans should check out the Bonsai Village where a collective of 14 miniature tree growers operate one of Japan's largest bonsai farms.

And if you're in Japan with the kids (or just a big kid yourself), you'll love the Dinosaur Exploration Hall featuring 250 dinosaurs computer-controlled to move just like the real thing!
More info: Saitama Prefectural Tourism Promotion Office. Tel: 048 830 3955

Getting to the game
Approximately 50 minutes from Tokyo via the Namboku subway line to Urawa-Misono Station on the Saitama railway line, then a 15 minute walk.

SAPPORO
Next, we will be travelling way up north to Sapporo, the capital of Hokkaido for the nerve-tingling match against Argentina.

Sapporo is one of Japan's largest cities with a population of over 1.8 million and there is plenty of fun to had in this unique city. Sapporo is known for its delicious local cuisine. Fish-lovers will warm to the Sampei-jiru one-pot stew of herring, cod and vegetables and the Genghis Khan barbecue (no, really) of lamb and vegetables is bound to go down well with Sven's barmy army.

The belt-like Odori park in central Sapporo is an oasis for residents and visitors alike and the Sapporo TV Tower at the eastern end of the park provides a panoramic view of the city. And fans wanting to lower their blood pressure further before the game are well-advised to take the hour's bus ride to the famous Jozankei resort and bathe in the Hot Springs.
More info: Sapporo Tourist Information Centre (at subway Odori Station): Tel. 011 232 7712

Getting to the game
A taxi from Sapporo Station will take half an hour. Alternatively, take the subway from Sapporo Station on the Toho Line to Fukuzumi Station (approx. 10 mins), then it's a 10-minute stroll to the 'Dome'.

OSAKA
By the time we meet Nigeria, Sven and the boys will already have qualified for the second round (well, hopefully!) so we fans can relax and explore Osaka, the largest city in Western Japan.

During the Edo period (1603–1868), Osaka boasted unrivalled prosperity as a trading hub for the whole country and it has continued to grow dynamically.

The Minami district is Osaka's leading amusement quarter, teeming with numerous eating and drinking spots and a huge variety of theatres (including Kabuki, Bunraku, modern theatrical and vaudeville).

Like Sapporo, Osaka will be a joy for fans with healthy appetites. Try local delicacies such as udonsuki (a casserole of noodles, seafood and vegetables), okonomi-yaki (thin flat cakes of batter fried with chunks of vegetables) or delicious Osaka-zushi (Osaka-style sushi).

If you prefer to watch fish swim rather than eat

TO YOKOHAMA

can discover about one of the world's most exciting countries...

them raw, visit the Kaiyukan aquarium in Osaka's waterfront leisure area, it is one of the world's largest aquariums containing 580 species of marine life from the Pacific Ocean.
More info: Osaka City Visitors Information Centre Tel. 06-6305-3311

Getting to the game
The Osaka World Cup stadium is a 5-minute walk from either the Tsurugaoka Station on JR's Hanwa Line or Nagai Station on the Midosuji subway line.

SECOND ROUND MATCHES

NIIGATA
If we finish second in our group, England's next match will be played in Niigata City. Dubbed the 'Waterside Capital', it is the largest population centre on the Sea of Japan coast.

A good place to start enjoying Niigata is the elegant Bandai Bridge crossing the Shinamo River that flows through the centre of the city. The bridge is a symbol of Niigata and there are centres either side renowned for their shopping and nightlife amenities.

Then, if you want to get a bit of air in your lungs after a night out (Niigata is known as the sake kingdom with 103 breweries), why not take a stroll along the Sasagawa-nagare, the breath-taking 11-km coastline along the Sea of Japan. Or you could take a day trip to nearby Sado Island. During the Edo period, the island thrived with gold mining and combined the samurai culture from Edo (now Tokyo) with the merchants' culture. These underlie the rich and unique culture of Sado where temples, historic sites and traditional performing arts such as Noh can be found aplenty. Bay cruises off Sado in glass-bottomed boats may also please seafaring fans.
More info: Niigata Prefectural Office Tourism Section tel. 025-285-5511

Getting to the game
From JR Niigata Station and Niigata Airport to the World Cup stadium takes approximately 10 and 25 minutes by car respectively.

OITA
After the hurly-burly of Osaka, if England win Group F, it's off to Japan's southernmost World Cup venue in the Oita Prefecture – and what a fantastic reward.

Oita is a place of such stunning natural beauty that it contains two national parks and three quasi-national parks. Known as the 'Land Of Abundance', the sea offers fresh fish and the mountains provide a stunning backdrop to the city. Add into the mix the Beppu Hot Spring resort and the hospitality of the Oita residents and you are sure of a warm welcome.

Must-visit attractions include the Natural Zoological Garden located on Mt Takasaki-yama midway between Oita and Beppu cities, where you can see some 2000 wild monkeys. And there are plenty of opportunities to dip into the rich cultural and spiritual history of the region. For instance, a visit to Usuki Sekibutsu – an array of 60 stone images of Buddha dating back 700 to 1200 years on a hill in the suburbs of Usuki City – is recommended.
More info: Oita Prefectural Tourist Association tel. 097-532-7305

Getting to the game
The stadium is not well-served by trains from the city centre, so take a taxi. From downtown Oita City to the to the stadium is about a 20-minute drive.

ROUGH GUIDE

QUARTER FINAL

SHIZUOKA

Now, if England win Group F and triumph in our second round match, we'll be going back to Osaka (see previous page). But if we started as runners-up, Shizuoka will host our quarter-final.

Only an hour from Tokyo by rail, many fans may opt to stay in the capital and just pop down to the stadium in Fukuroi City on matchday, but walkers and nature-lovers should pay a recreational visit.

Situated on the Pacific coast of the central Honshu region, Shizuoka Prefecture occupies the core section of the old Tokaido Road which has connected Tokyo and Kyoto since the Edo period. Although the road has since been replaced by the Tokaido Shinkansen super-express and the Tomei Expressway, Shizuoka's scenic splendour represented by Mount Fuji, the symbol of Japan, has remained unchanged. Mount Fuji is 3776 metres tall and can be viewed from various places around the prefecture. And the World Cup coincides with the July–August summer season when the summit is open to climbers of all ages, so pack some sturdy hiking boots if you fancy a bit of a challenge.

More info: Shizuoka Prefectural Tourist Association tel. 054-255-1388

Getting to the game

From Shizuoka City to the stadium in Fukuroi City takes approximately 50 minutes by JR Tokaido Honsen Line. By car, take the Tomei Expressway to Fukuroi IC.

SEMI-FINAL

Whatever route we take, if our lads make it to the semi-finals, we'll be back on familiar territory in Saitama where we played our first match against Sweden three weeks before. Computer-controlled dinosaurs, anyone?

THE FINAL

We made it! Yes, World Cup glory beckons for Sven and the boys and the magnificent 73,700-capacity stadium in Yokohama will be the stage for the final act.

Once a small fishing village, Yokohama opened to the Western world as one of Japan's first international ports in 1859. Since then it has blossomed into Japan's second largest city with a population of 3.4 million. Yokohama boasts Japan's tallest skyscraper, the 296-metre high Landmark Tower Building and the 'Sky Garden' on the 69th floor offers an amazing sweeping view from the heart of Tokyo to Mount Fuji.

Yokohama Chinatown is a popular spot for visitors. The second largest 'Chinatown' in the world in scale, over 500 establishments are clustered here, including restaurants offering all major styles of Chinese cuisine. But Yokohama can also satisfy those of you desperate for some European cooking after a month away. Due to the influence of the many foreigners who've settled here, a variety of cuisines are available, including Italian, French and Swedish to name but a few.

So that's Yokohama... now all we've got to do is get there!

More info: Tourist Information Centre, Yokohama Station tel. 045-441-7300

Getting to the game

From downtown Yokohama, take a ride on the JR Yokohama Line or the municipal subway from Sakaragicho Station to Shin-Yokohama Station. Either way, it's about a 15-minute journey then a 15-minute stroll to the stadium.

HANDY HINTS

Etiquette
It is polite to greet our Japanese hosts by saying 'How do you do' ('Haji-mem-ashi-te'), offering a handshake and bowing your head slightly. Most importantly, remember to take your shoes off whenever you go inside a building... It is very rude not to.

Accommodation
Save money by opting for one of the thousands of budget 'ryokans' in Japan instead of an hotel. If you shop around, a bed for the night could be available to you for as little as £15.

Travel
If you are planning to follow England from start to finish at World Cup 2002, it's well worth buying an unlimited use train ticket for £324. They are only available to buy abroad so buy in advance. Call the Japan Travel Centre on 020 7255 8283 for more details.

Food and drink
Generally, life isn't cheap in Japan. For instance, a burger will set you back £9 and a pint of beer is a tenner. But look out for local noodle bars where prices are much the same as they are back home in England.

TALK THE TALK

Impress our Japanese hosts with your grasp of the lingo using our handy phonetic phrase guide!

Hello
Kon-nichiwa
こんにちは

Goodbye
Sayounara
さようなら

How do you do?
Haji-mem-ashi-te?
はじめまして

I am glad to meet you.
Oaidekite ureshiideshu.
お会いできて嬉しいで

Shoot!
Shuuto!
シュート！

Goal!
Goooru!
ゴール！

There's only one David Beckham!
David Beckham wa hitori shika inai!
デービッド・ベッケナムは一人しかいない！

We are the champions my friend. We'll keep on fighting to the end.
Oretachiga champion sa. Saigo made tatakauzo.
俺たちがチャンピオンさ。最後まで戦うぞ。

Michael Owen is faster than a speeding bullet train.
Michael Owen wa shinkansen.
マイケル・オーウェンは新幹線より早い。

His name is Rio and he plays for Eng-er-land...
Kare no nawa Rio, igirisu no senshu da...
彼の名はリオ、イギリスの選手だ。

We're on our way to Yokohama...
Yokohama e ikuso...
横浜へ行くぞ。

THE VENUES

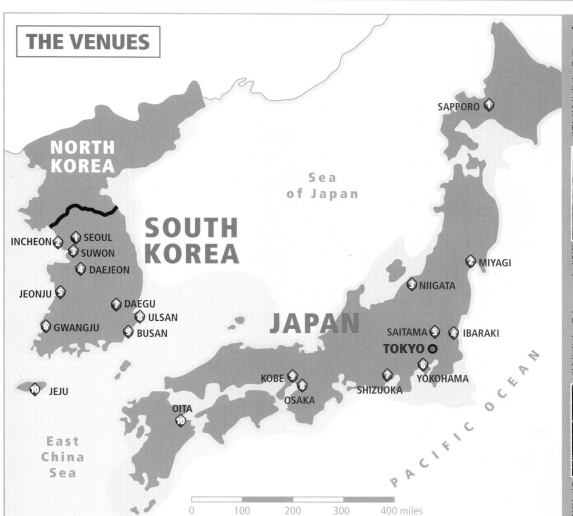

THE STADIA

Korea Republic

No.	STADIUM	CAPACITY
1	Seoul	63,930
2	Incheon	51,180
3	Suwon	44,047
4	Daejeon	42,176
5	Jeonju	42,477
6	Gwangju	42,880
7	Daegu	70,140

Daegu: Korea's largest stadium

8	Ulsan	43,003
9	Busan	55,982
10	Seogwipo	42,256

Japan

No.	STADIUM	CAPACITY
1	Sapporo Dome	42,122
2	Miyagi	41,800
3	Niigata	42,700
4	Ibaraki	49,133
5	Saitama	63,700

Saitama's super stadium

6	Yokohama	73,700
7	Shizuoka	51,349
8	Osaka	50,000
9	Kobe	42,000
10	Oita	50,000

ENDURING IMAGE

WONDER GOAL

30 June 1998, Saint Etienne
World Cup Finals Second Round
England 2 Argentina 2 (AET: Argentina won 4–3 on penalties)

Teenage prodigy Michael Owen dribbles past the entire Argentinian defence and clips the ball over goalkeeper Carlos Roa to give England a 2–1 lead. Unfortunately, Owen's world-class effort was eventually overshadowed by another painful defeat on penalties.

THREE LIONS

'We're playing for England... EN-GER-LAND!'

What have Jimmy Hill, Rod Hull & Emu and Shaun Ryder got in common? Why, they've all penned football songs for your listening, erm, pleasure...

FOUR YEARS OF EXPECTATION AND hope, but finally it's almost here. Yes folks, it's what we've all been waiting for... another World Cup song. Sorry, what do you mean you weren't waiting for it? Surely it's the highlight of the whole tournament (bar the odd Owen wonder goal), that time when the nation's footballing stars get together with the pop flavour of the month to record a singalong ditty for the benefit of the terrace crowds?

All The Way

Okay, okay, maybe football and recording studios aren't always the ideal marriage. But down the years the likes of *Here We Go* by the 1933 Arsenal squad and the more contemporary *Pass and Move (It's the Liverpool Groove)* have been sneakily bought in their hundreds of thousands to form the cornerstones of many people's record collections (for those under 15, records are the circular pieces of vinyl which you put on a turntable to play). Can you remember any of the following: *Ossie's Dream* by Chas and Dave, *Good Old Arsenal* (with lyrics by Jimmy Hill), *I'm Following Sheffield United* by Bobby Knutt and *All The Way* by Rod Hull and Emu?

World In Motion

Of course some songs earned their performers a bit of stick, especially the ones sung by (often duetting) footballers, such as *Where's The Ball* by Hurst and Peters, *Sugar Sugar* by Bobby Moore and Francis Lee, *Lily The Pink* by Jeff Astle and Peter Bonetti and, of course, *Head Over Heels* by Kevin Keegan.

However, some songs, like New Order's *World In Motion*, also earned their fair share of praise.

The Beautiful Game

In recent years, though, the modern pop star has been a bit more cunning with his football shenanigans. Albums such as *The Beautiful Game* featuring the likes of Supergrass, Black Grape and Teenage Fanclub have tried to give football a touch of contemporary 'cool'. Similarly, applying the modern touch in Scotland were Primal Scream who, in collaboration with *Trainspotting* author Irvine Welsh, produced *The Big Man And The Scream Team Meet The Barmy Army Uptown*. Nice.

England's Irie

Best of the bunch was probably *England's Irie*, a typically mad and inspired noise from Shaun Ryder's Black Grape. Ryder is known for some amazing lyrics, so it was unlikely that he would be fazed by the challenge of writing a football tune.

Vindaloo

France 98 saw the England team supported by *Top Of The World*, a song co-penned by Echo And The Bunnymen's Ian McCulloch and former Smiths guitarist Johnny Marr and featuring such luminaries as the Spice Girls and Simon 'Foxy' Fowler of Ocean Colour Scene. It was somewhat overshadowed by the vaguely lunatic offering from Fat Les: *Vindaloo* and the shadow still cast on all other footy tunes by *Three Lions*. The Skinner/Baddiel/Lightning Seeds number remains something of a terrace favourite. The 'Thirty years of hurt never stopped me dreaming' approach not only avoided the normal blind optimism of most football themes but also tapped into the feelings of hope, regret and nostalgia familiar to all football fans.

Top Of The Pops

At the time of going to press, we don't know what new joys await on the song front. But let's hope it's inspirational enough to see the boys all the way to the final in Yokohama.

'Three Lions' composers Baddiel & Skinner get in the box

WORLD CUP FUNNIES

by Tony Husband

THE LIKELY LADS

There is intense competition for places in England's World Cup squad. Let's take a closer look at the stars hoping to make it into Sven's final 22...

GOALKEEPERS

DAVID SEAMAN
CLUB ARSENAL
DATE OF BIRTH 19.9.63
HEIGHT 6FT 4IN
WEIGHT 14ST 10LB
CAPS 67 (INC. 4 SUB)
GOALS CONCEDED 38

England's most senior player, following the international retirement of Tony Adams, David Seaman has played for England under six different managers. He made his international debut in the friendly against Saudi Arabia in Riyadh in 1988 and his high point to date is probably those vital penalty saves during Euro '96. In addition, David has won two League Championships, two FA Cups, one League Cup and the European Cup Winners Cup in a distinguished club career with Arsenal.

NIGEL MARTYN
CLUB LEEDS
DATE OF BIRTH 11.8.66
HEIGHT 6FT 2IN
WEIGHT 14ST 10LB
CAPS 19 (INC. 4 SUB)
GOALS CONCEDED 15

Nigel Martyn became England's first '£1 million pound keeper' when he moved from Bristol Rovers to Crystal Palace in 1989 at the age of 23. When he moved on to Leeds five years ago he cost them £2.2 million. But despite the inflation, his England appearances total has not risen as rapidly as he would like, mainly because of David Seaman's consistency. Nigel won his first cap against the CIS in Moscow in 1992 and has had some excellent games of late for his country, particularly the vital qualifier against Greece where he made some crucial saves. Many critics' pick to wear the number one shirt.

DAVID JAMES
CLUB WEST HAM
DATE OF BIRTH 1.8.70
HEIGHT 6FT 5IN
WEIGHT 14ST 5LB
CAPS 5 (INC. 2 SUB)
GOALS CONCEDED 1

David James began his career with Watford before moving on to Liverpool, Aston Villa and now West Ham. A big, agile goalkeeper he has always shown signs of excellence and has now married that natural abilty to consistency. David won his first senior England cap against Mexico in 1997 and kept a clean sheet. A series of superb displays this season have gone a long way to silencing critics who once labelled him 'Calamity James'.

DEFENDERS

WES BROWN
CLUB MAN UNITED
DATE OF BIRTH 13.10.79
HEIGHT 6FT 1 IN
WEIGHT 12ST 11LB
CAPS 4 (INC. 2 SUB)
GOALS 0

An outstanding natural talent, Wes won his first cap at the age of just 19 in a friendly against Hungary in 1999. It was just reward for a series of mature displays during Manchester United's Treble-winning campaign. Wes then sustained a serious cruciate knee injury which kept him out of football for a year, but happily he has returned with his rich potential undiminished. A Manchester lad from the district of Longsight (hence the 'Longsight Libero' nickname bestowed on him by Reds' fans), Wes is renowned for his unflappable temperament and comfort on the ball.

SOL CAMPBELL
CLUB ARSENAL
DATE OF BIRTH 18.9.74
HEIGHT 6FT 2IN
WEIGHT 14ST 4LB
CAPS 42 (INC. 3 SUB)
GOALS 0

Since winning his first cap against Hungary in the build-up to Euro '96 Sol has established himself as a rock in the heart of the England defence. Sol's combination of immense physical strength and speed makes him exceptionally hard to beat and he is also capable of bringing the ball out of defence with authority – witness those mazy runs against Argentina at France '98! Last summer, after nine years of service to Tottenham Hotspur, Sol joined North London rivals Arsenal where he is seen as the natural successor to former England skipper Tony Adams.

ASHLEY COLE
CLUB ARSENAL
DATE OF BIRTH 20.12.80
HEIGHT 5FT 8IN
WEIGHT 10ST 8LB
CAPS 7 (NO SUB APPS)
GOALS 0

Left-back Ashley made the most of a loan spell at Crystal Palace before forcing Brazilian star Silvinho out of the Arsenal team in 2000–01 with mature displays that belied his youth. After just four caps at Under-21 level, he was fast-tracked into the senior side by Sven-Göran Eriksson and played in every World Cup qualifier since the 3–1 away victory over Albania last March. The 21-year-old Gunner is quickly learning the art of defending against international wingers and his natural left-footed crossing ability adds to England's attacking threat.

RIO FERDINAND
CLUB LEEDS UNITED
DATE OF BIRTH 7.11.78
HEIGHT 6FT 2IN
WEIGHT 12ST
CAPS 19 (INC. 5 SUB)
GOALS 0

Tagged 'the next Bobby Moore' as a teenager at West Ham, Rio is now showing real signs that he can justify such lofty comparisons. He made his debut for England against Cameroon in 1997, just eight days after his 19th birthday, and his potential as a stylish ball-playing central defender was immediately apparent. However, occasional lapses of concentration marred his generally assured displays and it has taken Rio a little time to establish himself at international level. He has thrived since his move to Leeds United, though, and he will be a key figure for England at Japan/Korea 2002.

MARTIN KEOWN
CLUB ARSENAL
DATE OF BIRTH 24.7.66
HEIGHT 6FT 1IN
WEIGHT 12ST 4LB
CAPS 40 (INC. 2 SUB)
GOALS 2

Born in 1966, Martin is one of the most senior members of the squad. He began his career at Arsenal, then had spells with Aston Villa and Everton before returning to the Gunners for £2 million in 1993. Martin made his England debut in a 2–0 friendly win over France in 1992, but he seems to have got better and better with age. A fierce competitor, Martin inspires the players around him with his drive and commitment and he was honoured with the captain's armband in England's second World Cup qualifier against Finland.

THE LIKELY LADS

GARY NEVILLE
CLUB MAN UNITED
DATE OF BIRTH 18.2.75
HEIGHT 5FT 11IN
WEIGHT 12ST 7LB
CAPS 49 (INC. 4 SUB)
GOALS 0

Gary has won five league championships, two FA Cups and a European Cup with Manchester United and is a veteran of two World Cup qualifying campaigns. That's a remarkable achievement and even more so when you consider that Bury-born Gary wasn't even picked for his county team as a teenager. His exponential rise to become an England regular at 20 was the result of dedication and his rigorous professionalism has made him England's premier right-back for the past seven years.

PHIL NEVILLE
CLUB MAN UNITED
DATE OF BIRTH 21.1.77
HEIGHT 5FT 11IN
WEIGHT 11ST 11LB
CAPS 34 (INC. 9 SUB)
GOALS 0

Like his brother, Phil has been part of a golden era of success at Manchester United. But Gary's domination of Phil's favoured right-back role has indirectly caused the younger Neville to fulfil a variety of other positions and his versatility is a useful asset for both Sir Alex Ferguson and Sven-Göran Eriksson. Phil made his senior debut in the final warm-up match prior to Euro '96, against China in Beijing and is determined to make the World Cup squad after just missing out in 1998.

GRAEME LE SAUX
CLUB CHELSEA
DATE OF BIRTH 17.10.68
HEIGHT 5FT 10IN
WEIGHT 12ST
CAPS 36 (INC. 3 SUB)
GOALS 1

The 33-year-old Chelsea left back has been overtaken by Ashley Cole at international level in recent times but shouldn't be written off. Jersey-born Graeme became Chelsea's joint club captain (alongside Gianfranco Zola) at the start of 2001–02 and has received rave notices for his play this season. A solid defender who can also be very dangerous going forward, Graeme could give the England manager more options in the left-back position and his experience of European Championship and World Cup tournaments is also in his favour.

GARETH SOUTHGATE
CLUB MIDDLESBROUGH
DATE OF BIRTH 3.9.70
HEIGHT 6FT
WEIGHT 12ST 6LB
CAPS 44 (INC. 9 SUB)
GOALS 1

Gareth came through the youth ranks at Crystal Palace and enjoyed six years in the Eagles' first team before moving on to Aston Villa in 1995. The same year, he made his England debut, coming on as substitute against Portugal at Wembley. He has produced many solid displays at international level since, recovering admirably following the penalty shoot-out in Euro '96. Now playing for Middlesbrough, Gareth will be pushing Rio and co hard for a place in the heart of England's defence at World Cup 2002.

MIDFIELDERS

DARREN ANDERTON
CLUB TOTTENHAM H
DATE OF BIRTH 3.3.72
HEIGHT 6FT 1IN
WEIGHT 12ST 5LB
CAPS 30 (INC. 2 SUB)
GOALS 7

Despite a series of injuries, Darren is an established member of the England squad. He was an ever-present at both Euro '96 and France '98 and could well play a part again this summer. A talented midfielder who can play in the centre or on the wing, Darren began his professional career at Portsmouth before signing for Spurs in 1992 where he has remained since. He made his England debut under Terry Venables in the Wembley friendly against Denmark in 1994.

NICK BARMBY
CLUB LIVERPOOL
DATE OF BIRTH 11.2.74
HEIGHT 5FT 7IN
WEIGHT 11ST 3LB
CAPS 23 (INC. 10 SUB)
GOALS 4

Nick Barmby has had a new lease of life since joining Liverpool in 2000, contributing to their revival in fortunes (not a fact that goes down too well with fans of his former club Everton). He began his career at Tottenham before being transferred for £5.25 million to Middlesbrough in 1995. He was also the first student of the F.A. National School to earn a senior cap, making his England debut at Wembley in 1995 against Uruguay.

DAVID BECKHAM
CLUB MAN UNITED
DATE OF BIRTH 2.5.75
HEIGHT 6FT
WEIGHT 11ST 9IN
CAPS 47 (INC. 3 SUB)
GOALS 6

London-born but a Manchester United fan all his life, David has been making headlines for years. He made his England debut in Glenn Hoddle's first match as coach, a World Cup qualifier against Moldova in Chisinau in 1996 and has gone on to become the England captain, and one of the most famous players in the world. But forget the hype, the most important aspect of Becks is that he is one of the finest footballers of his generation, as hard-working as he is gifted – and he doesn't take a bad free-kick either!

NICKY BUTT
CLUB MAN UNITED
DATE OF BIRTH 21.1.75
HEIGHT 5FT 10IN
WEIGHT 11ST 11LB
CAPS 15 (INC. 7 SUB)
GOALS 0

A tough, competitive, but frequently under-rated midfielder, Nicky is another graduate of Manchester United's endlessly productive School Of Excellence. Despite a midfield that includes the likes of David Beckham, Ryan Giggs, Roy Keane and Juan Sebastian Veron, Butt has managed to earn a regular slot in the United first team and been a regular squad member for England. He made his international debut in 1997, coming on as a substitute for Steve McManaman in the 2–0 friendly win over Mexico at Wembley.

JAMIE CARRAGHER
CLUB LIVERPOOL
DATE OF BIRTH 28.1.78
HEIGHT 6FT 1IN
WEIGHT 13ST
CAPS 7 (INC. 5 SUB)
GOALS 0

Jamie Carragher has made over 100 appearances for Liverpool, the club he has been with since he was a teenager. His versatility has been a particular asset for the team and he is able to play in a variety of defensive and midfield roles for the Anfield club. Jamie established a new record for England Under-21 appearances (27) before stepping up to the senior squad in 1999. He made his first appearance coming on as a sub against Hungary in Budapest.

KIERON DYER
CLUB NEWCASTLE
DATE OF BIRTH 29.12.78
HEIGHT 5FT 8IN
WEIGHT 10ST 1LB
CAPS 8 (INC. 4 SUB)
GOALS 0

Initially a trainee at Ipswich Town, Kieron Dyer graduated to the first team and soon established himself as one of the club's best players. It was inevitable that other clubs would start to show an interest in the versatile midfielder and it was Newcastle who got his signature. Since arriving at St James' Park, Kieron's progress has continued apace. He made his England debut in the European Championship qualifier against Luxembourg at Wembley where his attacking flair contributed to a 6–0 victory.

STEVEN GERRARD
CLUB LIVERPOOL
DATE OF BIRTH 30.5.80
HEIGHT 6FT 2IN
WEIGHT 12ST 4LB
CAPS 8 (INC. 1 SUB)
GOALS 1

A key player for Liverpool and England, Steven Gerrard still seems to be improving with every game. Coming through the Liverpool ranks, he has established himself as one of the finest young midfielders in Europe and was a triple Cup-winner last season. For England, he produced a moment of magic to score a superb goal against the Germans in the 5–1 away victory. He made his senior international debut in May 2000 against Ukraine and, injuries permitting, looks set to be an England regular for many years to come.

OWEN HARGREAVES
CLUB BAYERN MUNICH
DATE OF BIRTH 20.1.81
HEIGHT 6FT 1IN
WEIGHT 13ST
CAPS 2 (INC. 1 SUB)
GOALS 0

Already the holder of a Champions League winners' medal, Owen was snapped up by Bayern Munich at the age of 16 when they saw him playing in Canada. Thanks to his parentage and career he was actually eligible to play for three countries, so it was great news when Owen plumped for his father's native land. With two caps already (against Holland and Germany), this skilful player is certain to feature in future England teams.

THE LIKELY LADS

STEVE McMANAMAN
CLUB REAL MADRID
DATE OF BIRTH 11.2.72
HEIGHT 6FT
WEIGHT 10ST 6LB
CAPS 37 (INC. 12 SUB)
GOALS 3

As a local lad, Steve McManaman was a Kop favourite at Liverpool for whom he played over 350 games, before joining Real Madrid in 1999. He has thrived in Spanish football and played a vital role in Real's 2000 European Cup triumph. Steve made his England debut in 1994 as sub in a friendly against Nigeria and he started all of England's matches at Euro '96.

DANNY MURPHY
CLUB LIVERPOOL
DATE OF BIRTH 18.3.77
HEIGHT 5FT 9IN
WEIGHT 10ST 8LB
CAPS 1 (INC. 1 SUB)
GOALS 0

Liverpool paid Crewe £1.5 million to bring this versatile midfielder to Anfield in 1997 and by the 2000–01 Treble season he had established himself in the Reds' first team. Danny had already won five Under-21 caps before making his senior debut as a substitute against Sweden last year.

PAUL SCHOLES
CLUB MAN UNITED
DATE OF BIRTH 16.11.74
HEIGHT 5FT 7IN
WEIGHT 11ST 8LB
CAPS 40 (INC. 2 SUB)
GOALS 13

As well as winning trophies for Manchester United, Paul has been a pivotal member of the England squad for the last three managers. Consistent and spectacular, Paul's goals have often saved England's blushes and he made a real contribution in the team's qualifying matches with goals against Albania and Greece away from home. There's no doubt he can be relied upon to deliver more heroic displays at World Cup 2002.

TREVOR SINCLAIR
CLUB WEST HAM
DATE OF BIRTH 2.3.73
HEIGHT 5FT 10IN
WEIGHT 12ST 10LB
CAPS 1 (NO SUB APPS)
GOALS 0

Another product of the F.A. National School, Trevor was a member of the England squad from his days at QPR. But it wasn't until last November that he finally made his senior international debut, in a friendly against Sweden at Old Trafford. A fine midfielder with a penchant for scoring spectacular goals.

FORWARDS

ANDREW COLE
CLUB BLACKBURN
DATE OF BIRTH 15.10.71
HEIGHT 5FT 10IN
WEIGHT 12ST 4LB
CAPS 15 (INC. 6 SUB)
GOALS 1

Andrew Cole is Manchester United's all-time top scorer in European competition with 18 strikes but has found international goals harder to come by. He had to wait six years before breaking his duck against Albania last March. A superb all-round striker, Andrew has always proved his critics wrong and played an important role in England's qualifying campaign.

ROBBIE FOWLER
CLUB LEEDS UNITED
DATE OF BIRTH 9.4.75
HEIGHT 5FT 11IN
WEIGHT 11ST 10LB
CAPS 22 (INC. 11 SUB)
GOALS 5

In December 2001, Robbie made the momentous decision to leave Liverpool and join Leeds United after a decade at his hometown club. Robbie is arguably the most naturally-gifted finisher in England and expected to be on the plane to Japan. He won his first cap against Bulgaria in 1996 and notched his first England goal against Mexico a year later. He has now scored five, including a brilliant individual effort against Albania last September.

EMILE HESKEY
CLUB LIVERPOOL
DATE OF BIRTH 11.1.78
HEIGHT 6FT 2IN
WEIGHT 13ST 12LB
CAPS 20 (INC. 11 SUB)
GOALS 3

The £11-million Liverpool striker enhanced his chances of partnering club-mate Michael Owen for England at World Cup 2002 with his excellent contribution to the historic 5–1 win over Germany. Emile was rewarded with a starting place up front for the last two qualifiers against Albania and Greece. But Sven-Göran Eriksson has also employed Emile in the left-wing role during the campaign, where the former Leicester star's pace and power can unnerve opposing defenders.

MICHAEL OWEN
CLUB LIVERPOOL
DATE OF BIRTH 14.12.79
HEIGHT 5FT 8IN
WEIGHT 11ST
CAPS 32 (INC. 10 SUB)
GOALS 14

The 2001 European Footballer Of The Year's exceptional acceleration and icy finishing skills helped him to a six-goal tally in the qualifiers. Now the youngest English international of the twentieth century (he made his debut against Chile in 1998 aged 18 years and 59 days) appears a certainty to spearhead England's attack at World Cup 2002 and Brazilian legend Pele reckons he will 'need the full-time attention of at least two defenders'.

TEDDY SHERINGHAM
CLUB TOTTENHAM H
DATE OF BIRTH 2.4.66
HEIGHT 5FT 11IN
WEIGHT 12ST 5LB
CAPS 43 (INC. 13 SUB)
GOALS 11

One half of the famous 'SAS' partnership with Alan Shearer at Euro '96, 36-year-old Teddy is still regarded by many pundits as the best support striker in England. He made his international bow in a World Cup qualifier against Poland in Chorzow in 1993 and has scored 11 and set up even more goals in a distinguished international career. Teddy was a key figure in Manchester United's historic Treble win in 1998–99 and last summer he returned to Tottenham Hotspur where his exceptional form has continued.

ALAN SMITH
CLUB LEEDS UNITED
DATE OF BIRTH 28.10.80
HEIGHT 5FT 9IN
WEIGHT 11ST 10LB
CAPS 3 (INC. 3 SUB)
GOALS 0

The fiery young striker was rewarded for a fine season for Leeds United with his first senior cap in the Mexico friendly in May 2001, coming on as sub for Michael Owen. Alan is a product of Leeds United's youth policy and the Elland Road faithful have come to admire his combative style and never-say-die attitude. He has already demonstrated his ability to perform at the top level, netting several valuable goals during Leeds' run to the semi-finals of the 2000–01 Champions League.

THE HOPEFULS

On top of the 30 players profiled already, there are plenty more Premiership stars in with a shout of going to Korea/Japan 2002.

Arsenal new boy Richard Wright is pressing hard for one of the three goalkeeper slots in the squad. Now 24 years old, Richard won 15 caps for the Under-21s and has been involved with the senior squad since 1998. The experienced Leicester 'keeper Ian Walker will also have high hopes of making the squad.

In defence, the accomplished Charlton left-back Chris Powell has shown that age is no barrier to the international scene since making his England debut at 31 last February against Spain. At the other end of the scale, Aston Villa left back Gareth Barry, who first joined the England squad back in February 2000, will be looking to force his way back into the reckoning. On the right-hand side, Leeds full-back Danny Mills could challenge Gary Neville, while promising Chelsea central defender John Terry and Middlesbrough's Ugo Ehiogu will bid for central defensive places.

The 20-year-old West Ham midfield starlet Joe Cole is regarded as a 'special talent' by Sven-Göran Eriksson and a consistent 2001–02 season for Cole and his club-mate Michael Carrick could earn them World Cup berths. Smooth-passing Frank Lampard will be hoping his move from Upton Park to Chelsea can kick-start his international career and hard-working Arsenal stalwart Ray Parlour is also in the frame. Meanwhile, Blackburn's David Dunn is widely-tipped to make the jump from England Under-21 to senior level and fill the troublesome left midfield role.

In attack, the finishing skills of Sunderland's 28-year-old Kevin Phillips are hard to ignore while Bolton striker Michael Ricketts has revealed a big-game appetite in his first Premiership season.

So who will make it and who will be disappointed? Sven, it's over to you...

PICTURE CREDITS

The publishers would like to thank the following sources for their kind permission to reproduce the pictures in this book:

Action Images; Allsport UK Ltd; Empics; FAOPL; Hulton Archive; Korea National Tourism Organisation; Japan National Tourist Organization; Popperfoto

Every effort has been made to acknowledge correctly and contact the source and/or copyright holder of each picture, and Carlton Books Limited apologises for any unintentional errors or omissions which will be corrected in future editions of this book.

The publishers wish to express their sincere gratitude to Simon Mooney and everyone at FAOPL for the time and assistance they have contributed to this project.